# Deep Swarm and Evolution for Generative Artificial Intelligence

**Hitoshi Iba**
Professor, Graduate School of Information Science and Technology
University of Tokyo, Tokyo, Japan

CRC Press
Taylor & Francis Group
Boca Raton London New York

CRC Press is an imprint of the
Taylor & Francis Group, an **informa** business
A SCIENCE PUBLISHERS BOOK

First edition published 2025
by CRC Press
2385 NW Executive Center Drive, Suite 320, Boca Raton FL 33431

and by CRC Press
4 Park Square, Milton Park, Abingdon, Oxon, OX14 4RN



*CRC Press is an imprint of Taylor & Francis Group, LLC*



*Library of Congress Cataloging-in-Publication Data (applied for)*

ISBN: 978-1-032-81577-0 (hbk)
ISBN: 978-1-032-82807-7 (pbk)
ISBN: 978-1-003-50633-1 (ebk)

DOI: 10.1201/9781003506331

Typeset in Times New Roman
by Prime Publishing Services

# Preface

> Early AI was mainly based on logic. You're trying to make computers that reason like people. The second route is from biology: You're trying to make computers that can perceive and act and adapt like animals. (Geoffrey Hinton, 2011)

This book provides theoretical and practical knowledge about swarm and evolutionary approach, i.e., deep swarm and deep evolution, to generative AI. While the central theme of the book is generative AI, it also develops a discussion of AI in a broader sense. In particular, we will discuss how the "emergence" concept can contribute to the improvement of AI.

Including my student days, I have been conducting research on AI for more than 30 years. The recent rise of AI is truly remarkable. I have been giving lectures on AI at universities for almost 20 years. Recently I am often troubled by the fact that young students simply want to hear about the latest AI applications.

AI has had two winters, and I have experienced it since the first winter. The first spring, however, I only know about through rumors from my senior researchers. Now we are in the third spring. Deep learning, machine learning, and generative AI are at the center of it.

In my AI lectures, I used to tell my students to get Ig Nobel Prize as a joke, since Nobel Prize is impossible in computer science. In fact, some of my fellow researchers have won Ig Nobel Prize several times. I also offered the topic of Foldit! in relation to my own past research in genome science.

However, while I was writing this book, the Nobel Prize was announced. The physics and chemistry prizes are related to AI.

I knew Geoffrey Hinton, one of the winners, from PDP (Parallel Distributed Processing) book that I read as a student in 1985. Hinton has consistently engaged in neural network research since then and endured the above-mentioned two winters. His attitude toward research is truly admirable. Regarding the

Baldwin effect described in this book, Hinton wrote a paper on the co-adaptation in case of the "needle in a haystack" problem nearly 40 years ago.

Recently I have found that young students seem to be less focused on one thing and less patient in their research. This may be because society demands too much in the way of earlier results. In the past, profound research that would bear fruit in ten years or more was acceptable. Philosophy and conviction in research were also considered essential.

I have been studying evolution and emergence since graduate school. In this field, it is essential to relate biological organisms and AI. Therefore, my desire to be close to evolutionary adaptive bio-systems has not changed as an underwater naturalist. In this book, I have included photographs related to evolution, such as those taken in Galapagos islands and Japanese diving sites (Iriomote and Amami). This is because encounters with these creatures have given me hints for the research on emergence and swarm intelligence.

Evolution is a slow process. Therefore, I hope that this book will provide the readers with a research agenda for the future, learning from the past.

There is no guarantee that a third winter will never come.

# Acknowledgments

I am deeply grateful to all of these wonderful people for helping me finish the project of this book. In particular, I would like to thank my colleagues and students in my laboratory at the Graduate School of Information Science and Technology of the University of Tokyo for creating a comfortable research environment.

This book is based on the class notes of "Artificial Intelligence," "Systems Engineering," and "Simulation Theory" that I used at the university. Many of the assignments given in class were quirky and often seemed difficult or unanswerable. However, the reports that were returned often contained impressive descriptions and deep insights, for which I was always grateful. In this book, I have used some of the sample answers and opinions from the submitted reports, with necessary modifications. Although it is impossible to name everyone, I would like to express my deepest gratitude to all the students who made the efforts and wrote interesting reports.

I would also like to express my deepest gratitude to SmartRams, Inc. for providing the design image for the cover of this book.

And finally, I would like to thank my wife Yumiko, and my sons and daughters Koki, Hirono, and Hiroto, for their patience and cooperation.

**Hitoshi Iba**

March, 2025

Okinawa, Japan

# Contents

# List of Figures

# List of Tables

# Chapter 1

# Introduction

Evolution is cleverer than you are (Orgel's Second Rule attributed to Francis Crick).

## 1.1 Evolution at work

Consider the following questions:

- Why do fish school together? (see 1.1(a))
- Why are peacocks so beautiful?
- Why do working ants serve their queen without having children themselves?
- Why do fish and shrimp cooperate? (see 1.1(b))

Solving these riddles leads us to the concept of evolution.

Let us consider what is necessary for evolution. Obviously, evolution requires more than a single animal (person). In other words, evolution requires a group or a population. Charles Darwin[1] found that populations of organisms must have the following characteristics for the sake of evolution:

---

1 
Charles Darwin (1809–1882): British natural scientist. Outstanding geologist and biologist. In 1858, he and Russell Wallace (see also page 35) published their theory of evolution, and the following year Darwin published "The Origin of Species."

(a) A school of hammerhead sharks (Mikomoto, Japan; July, 2024).

(b) Two yellow prawn-gobies living together in a hole made by a snapping shrimp in the sand (Iriomote, Japan; September, 2024).

**Figure 1.1:** Why do fish cooperate with each other?

- The members must be capable of producing offspring (i.e., self-reproduction). If unconditional, they produce an excess of offspring.
- The offspring must carry forward the characteristics of the parent(s), with some changes (i.e., variation).
- The offspring that have adapted to the environment must survive with a higher probability (i.e., survival of the fittest).

The first characteristic of overproduction is based on the principle developed by Thomas Malthus in his "An Essay on the Principle of Population." It states that population increases geometrically (exponentially) if not limited, but living resources increase only arithmetically. Darwin is said to have arrived at his theory of evolution after reading this. From these three characteristics, if adaptive traits are inherited to any degree, the advantageous traits will be passed on to the next generation.

The goal of evolutionary methods is to realize computational systems (evolutionary systems) based on this idea. The idea of evolution is not limited to the natural world, but is everywhere. Take, for example, the airplane, one of mankind's greatest inventions. Wright Brothers, the inventors of the Wright Flyer, went through many trials and errors before they succeeded in taking off. After repeated failures in the development process, they tried minor changes (= mutations), drawing out the best ideas from the past. We also combined good ideas into the next prototype (i.e., crossover). The process of repeatedly making prototypes and improving their accuracy is the very work of an evolving organism. Humans have also unconsciously used evolutionary mechanisms in the breeding of livestock and racehorses since ancient times. Evolutionary computation uses the same mechanism.

The basic data structures in evolutionary computation are based on knowledge of genetics. Hereafter, we shall provide an explanation of these.

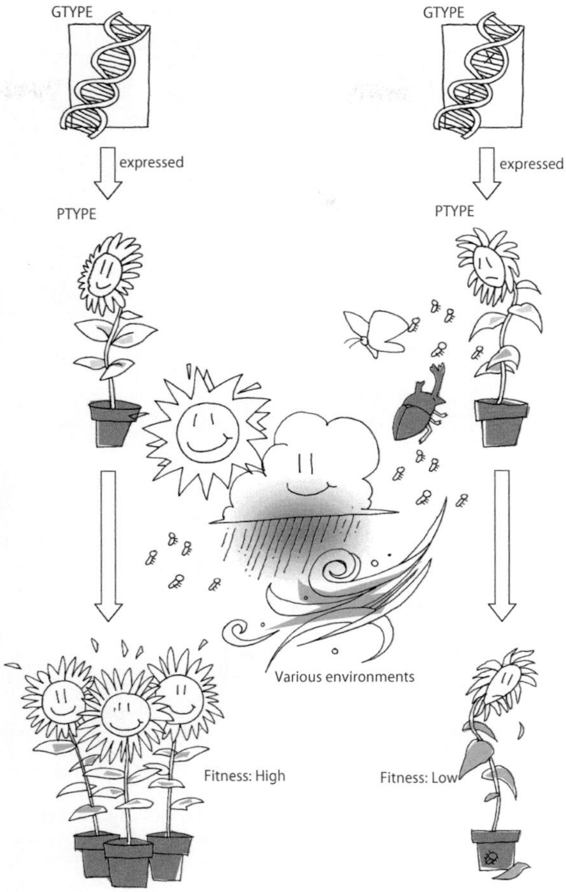


**Figure 1.2:** GTYPE and PTYPE.

The data used in evolutionary computation are formed from the two-layer structures of PTYPE and GTYPE. GTYPE (genotype, also called genetic codes) are, in a genetic type analogy, a low-level, locally-regulating set. This is the evolutionary computation to be operated on, as described later. The PTYPE is a phenotype, and expresses the emergence of behavior and structures over a wide area, accompanied by development within a GTYPE environment. Fitness is determined by the PTYPE adapting to its environment, and selection relies on the fitness of the PTYPE (see Fig. 1.2). For a time, the higher the fitness score taken, the better. Therefore, for individuals with a fitness of 1.0 and 0.3, the former can adapt better to their environment, and it is easier for them to survive (however, in other areas of this book, there are cases when it is better to have a smaller score).

We shall explain the basic framework of the evolution computation, based on the above description (see Fig. 1.3). Here, we configure a population containing

**Figure 1.3:** Image of evolutionary computation.

several dogs. We shall call this generation $t$. This dog has a genetic code for each GTYPE, and its fitness is determined according to the generated PTYPE. In the diagram, the fitness of each dog is shown as the value near the dog (remember that the larger the better). These dogs reproduce and create the descendants in the next generation $t+1$. In terms of reproduction, the better (higher) the fitness, the more descendants they are able to create, and the worse (lower) the fitness, the more easily they become extinct (in biological terminology, this refers to selection). In the diagram, the dog fur patterns undergoing slight change in the phenotypes due to reproduction are drawn schematically. As a result of this, the fitness of each individual in the following generation $t+1$ is expected to be better than that of the previous generation. Furthermore, the fitness as seen in the population as a whole also increases. In the same way, the dogs in the generation $t+1$ become parents and produce the descendants in the generation $t+2$. As this is repeated and the generations progress, the whole population improves.

In the case of reproduction, the operator shown in Fig. 1.4 is applied to the GTYPE, and produces the next generation of GTYPEs. For the sake of simplicity,

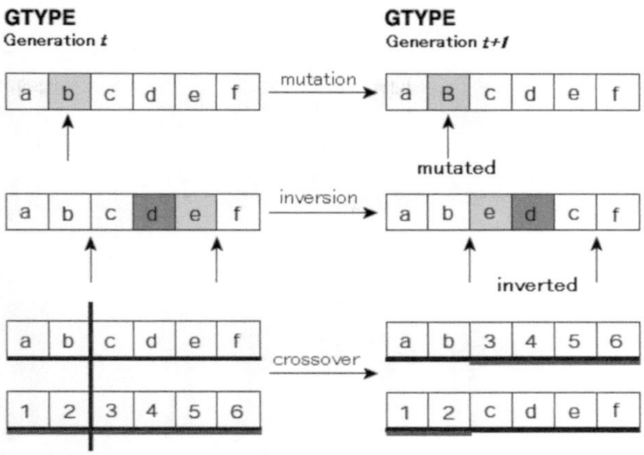


**Figure 1.4:** Genetic operators.

here, the GTYPE is expressed as a 1-dimensional matrix. Each operator is an analogy for the genetic recombination and mutation, etc., in the organism. The application frequency and the application area of these operators are randomly determined in general.

Normally, the following kinds of methods are used for selection:

- **Roulette selection**: This is a method of selecting individuals in a ratio proportionate to their fitness. A roulette is created with an area proportionate to fitness. The roulette is turned and the individual in the area where its needle stops is selected.
- **Tournament selection**: A certain number (called the tournament size) of individuals are randomly selected. The individual with the best fitness is chosen from among these individuals. This process is repeated for the same times as the size of the population.
- **Elite strategy**: Some better fitness individuals are left intact in the next generation. This prevents better individuals from being eliminated. This strategy is used in combination with the above two methods.

To summarize, in the evolutionary computation, generational change is shown in Fig. 1.5. In the figure, $G$ is the elite rate (rate of upper level individuals that were copied and left results). We can refer to the reproduction rate as $1 - G$.

Let us see how useful evolutionary computation can be in a simple spider simulation. The spider spins a radial thread and then builds a spiral thread. In this simulation the spider's genes are supposed to contain information that defines the radial angle and helical distance for building a web. What shape of web should be built? This can be measured by the performance of the nest. Naturally, the

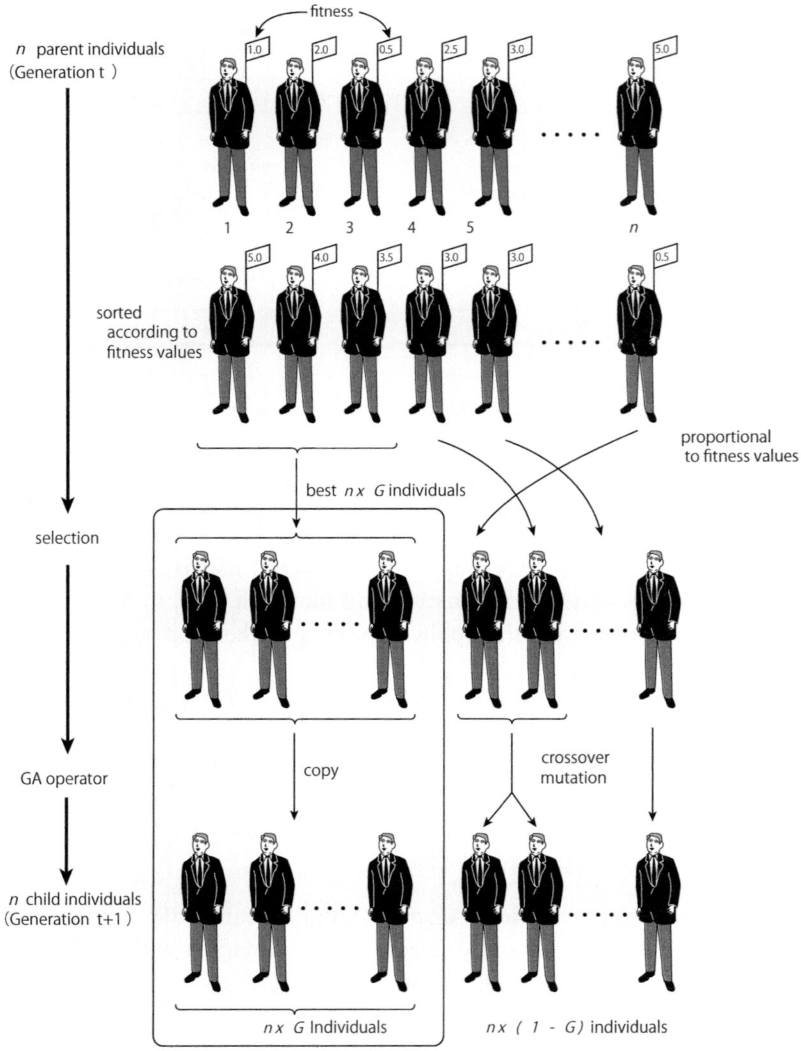


**Figure 1.5:** Selection and reproduction of GA.

purpose of a nest is to catch as many flies as possible. On the other hand, a nest that is too large or dense will cost too much because of the huge amount of thread used. Therefore, the degree of suitability should be defined in consideration of both of these factors. In details, the following fitness is used for evolution:

$$\text{Fitness of a nest} = \frac{\text{Number of insects caught within the nest}}{\text{Total amount of threads spun for the nest}} \tag{1.1}$$

To determine the number of insects caught, we randomly fly insects approximated by a small circle to see if they come in contact with the nest thread.

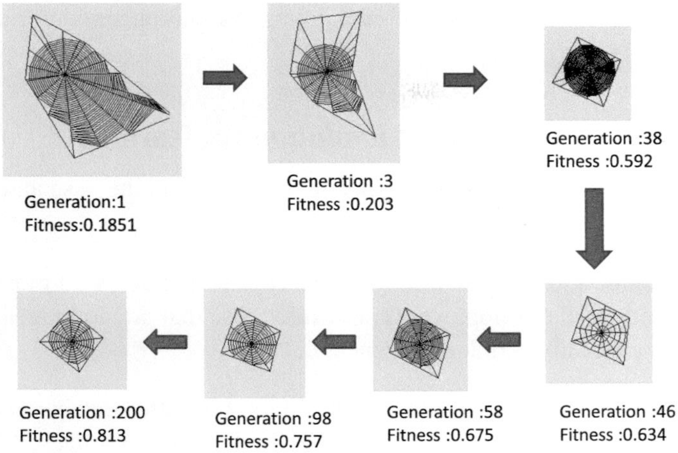


**Figure 1.6:** Evolution of a spider web nest.

Figure 1.6 shows the shape and fitness of web nests resulting from evolution. Initially, asymmetry and overuse of threads are observed, but eventually symmetrical and efficient nests are formed. This experiment was simulated with a population size of 1,000, a mutation rate of 0.1, and an elite rate of 0.1. The number of insects flown at random is 400.

In his 1986 book *The Blind Watchmaker* [27], Richard Dawkins[2] describes how images created according to simple rules can evolve into extremely complex and intriguing ones as a result of mutation and user selection. These image sequences were called *Biomorphs*, since they appear to resemble animals (e.g., insects) at first sight (see Fig. 2.8 in in Section 2.2.1).

If a stone falls in the forest, there is no question that it is there. But suppose that it is a watch. In that case, one would assume that someone dropped it. In other words, there must be something (the watchmaker) with the intention of making and placing the watch. Similarly, since living organisms (their bodies and functions) are elaborately optimized, it is tempting to think that living things were designed by someone (God?) with a clear intention. This idea is called "natural theology." But Dawkins and other evolutionists have vigorously argued that natural selection played the role of watchmaker. Natural selection is a blind watchmaker, so to speak, having neither vision nor intent. However, over time, even complex biological structures such as the eye have been shown to evolve.

---

2  Richard Dawkins (1941–): English evolutionary biologist, who has written many general books and general introductions to biology and, as a result, espoused thinking on the selfish gene, meme (cultural information replicators), and expanded phenotypes (host operation due to the parasites, dams made by beavers, and mounds of white ants can be seen as phenotypes). His revolutionary ideas and provocative comments about evolution are still causing many discussions. He is a famous atheist.

We can see an effective aspect of evolution in Biomorph simulator described in Section 2.2.1.

### 1.1.1 Nature or nurture: Evolution v.s. learning

Lamarck[3] proposed the following theory. This is called the "use and disuse theory."

> Depending on the environment, the organs that are used most frequently will develop over time, while those that are no longer used will gradually shrink or degenerate.

A famous example is a giraffe (see Fig. 1.7). A giraffe stretches out its neck in search of young leaves to feed on at increasingly higher altitudes. As a result of this behavior being repeated over a long period of time, the forelimbs and neck of the giraffe gradually lengthened. Lamarck's theory of evolution is based on the "inheritance of acquired traits." However, with the subsequent development of genetics, this theory was completely rejected. Although Lamarck is second only to Darwin as the most important biologist who clearly considered the fact that organisms evolve, his theory itself was not correct. Nonetheless, Lamarck's theory has been considered by many scholars. This may mean that the idea of inheritance of acquired traits appeals to our intuition.

Even if the acquired traits themselves are not inherited, it is hard to imagine that learning, which is an important factor for organisms, is completely unrelated to evolution (heredity). In the following, the relationship between learning and heredity will be explained with the Baldwin effect, and the application to evolutionary computation will be described in Section 2.5.

In 1896, Baldwinian evolution and the Baldwin effect were proposed independently by James Baldwin[4] [7], Morgan [88] and Orsborn [95], also known as ontogenic evolution. Later in 1942, Waddington [154] reviewed the theory and presented the term "genetic assimilation" or "canalization." In this theory, acquired traits cannot transfer phenotypic information directly back to the genotype. Instead, when an individual learns good traits in its lifetime, it is rewarded in the selection. Therefore, the selection priority is automatically assessed to the genotypes, inherently easier to acquire good traits. It is denoted that the children of a neck-stretching animal do not necessarily receive long necks, but the good ability in the stretching. The children of such a parent will be excellent in various activities (see Fig. 1.7). The Baldwin effect has always been controversial in biology, because in the complex natural world it is too difficult to verify it.

---

[3]Jean-Baptiste Pierre Antoine de Monet, Chevalier de Lamarck (17441–1829): Famous 19th century naturalist. He was one of the first people to use the word "biology" in its modern sense.

[4]James Mark Baldwin (1861–1934): American philosopher and psychologist.

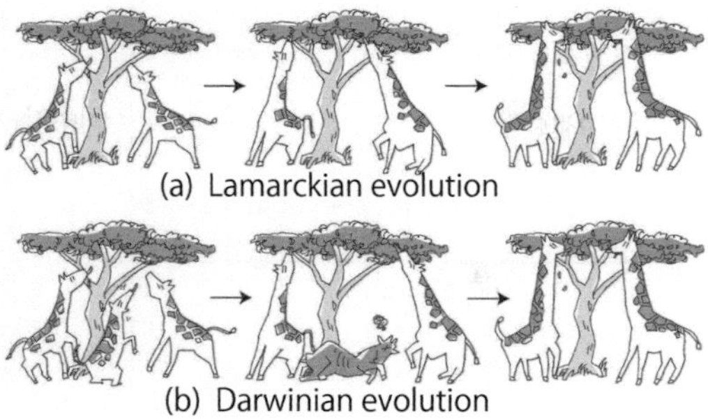


**Figure 1.7:** Different types of evolution.

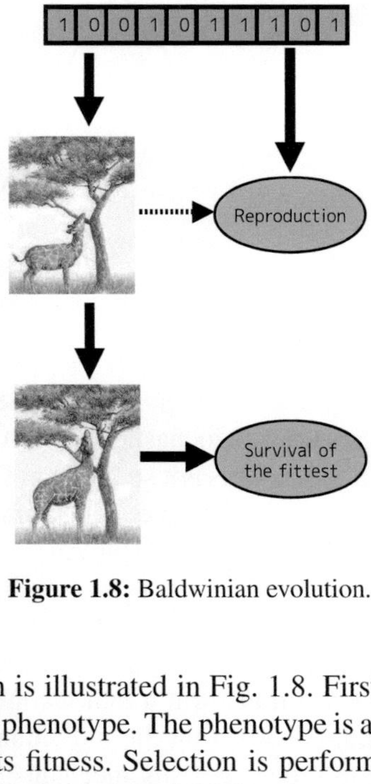


**Figure 1.8:** Baldwinian evolution.

Baldwinian evolution is illustrated in Fig. 1.8. First the individual translates its initial genotype to the phenotype. The phenotype is adapted through a learning process, and improves its fitness. Selection is performed based on the adapted fitness. Mating is performed using the initial genotype. While the traits acquired during learning are not transferred back to the genotype, it has the potential to develop good or bad final phenotypes based on learning.

For many problems, the search for an optimum is similar to finding a needle in a haystack. There will be an acute fitness decrease when leaving the optimum, such as shown in Fig. 1.9. Then, even if an individual is near a local optimum, it will not be able to detect the local optimum and thus will not find the global optimum.

If the evolving individuals can learn, so that they can search in a neighborhood around its initial position, the ability to find a local optimum will be

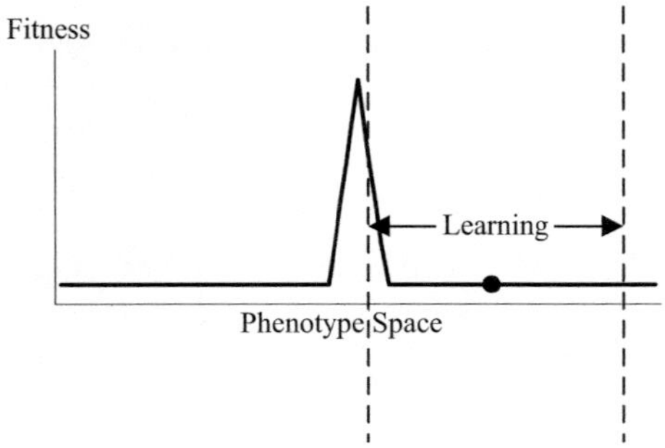


**Figure 1.9:** Learning enables more individuals to find a needle.

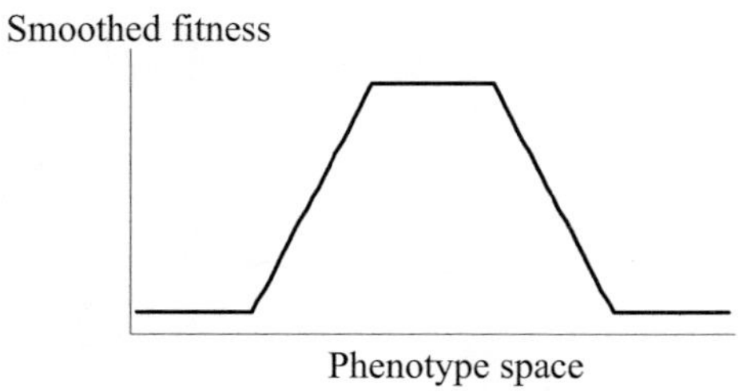


**Figure 1.10:** Equivalent fitness landscape smoothed by learning.

broadened. The learning process can be implemented by constructing an analytic model of the local area, by a stochastic search method, or by some random trials. In this case, only if the point is not so far from the spike, it may be possible to reach the spike by learning, i.e., by receiving some fitness bonus and calling attention to the area (see Fig. 1.9). With learning, it is possible to search in an equivalently smoothed landscape, just as shown in the curve in Fig. 1.10.

However, learning always incurs in a cost. This is true of evolutionary computation as well as in the natural world. If this learning cost is included in the selection mechanism, we can transform it in a second drive for approaching the optimum [145]. More traits learned by the individual will become fixed to the genotype as the evolution proceeds. This mechanism is named as "genetic assimilation," or "canalization" [134].

By chance, possibly affected by mutation, certain genius individuals may be born with good traits that its ancestors spent much time to learn. These individuals have a higher starting point, thus cost less to learn what their ancestors learned, Under the selection pressure, such genius individuals will win in the selection and the species is allowed to obtain even better phenotypes as a result of learning. As shown in Fig. 1.11, we can reach the top of the hill by learning form both points A and B. However, due to the low cost in learning, individuals starting from B will beat those from A in the selection. Although Baldwinian

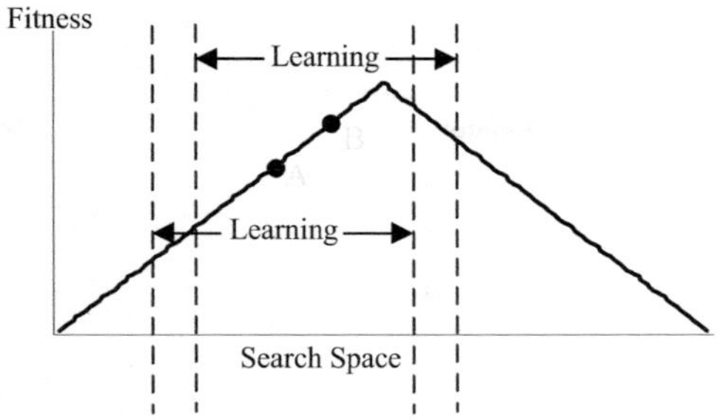


**Figure 1.11:** Genetic assimilation.

evolution does not transfer acquired traits back to the genotype, it can be observed that the evolution follows the indication of learning. Waddington [154] noted the process by which a phenotypic character becomes taken over by some genotype through evolutionary selection. This process would begin even in the absence of the environmental influence which was at first necessary.

Affected by these two basic mechanisms, the Baldwin effect acts as a two-phase process in the search (see Fig. 1.12). As a whole, the learning cost is adapted by the selection pressure automatically. The two phases are described as follows:

**Phase1** Individuals which have gained an advantage from a learned adaptive trait proliferate throughout the population.

**Phase2** Individuals with a lower cost of learning, i.e., those which innately acquire the adaptive trait, proliferate throughout the population.

In the first phase, the individuals do not fit the environment well when directly determined by genotypes. The individuals with more potentials in learning, or more parts left to learning, are likely to attain higher fitness in its lifetime. The learning cost grows, but not as fast as the increase of the fitness. As a result, such individuals will survive, and the learning-assisted search may dominate the whole population. This search explores different areas in the search space and the fitness grows up remarkably, while learning costs rise and keep high.

In the second phase called "genetic assimilation," most individuals can reach a similar level of fitness after lifetime learning. At this stage, increasing cost in learning helps little in improving final fitness, while more punishment is imposed because of such high cost. As a result, individuals with less learning costs to reach the optima will survive so that the genetic assimilation will dominate. This search concentrates on the exploitation and the fitness increases steadily, while learning costs decrease, even to zero in some situations.

Evolutionary computation methodologies also vary in implementation depending on whether they are based on Darwinian, Lamarckian, or Baldwinian theories. For Darwinian evolution, natural selection is performed after evaluating

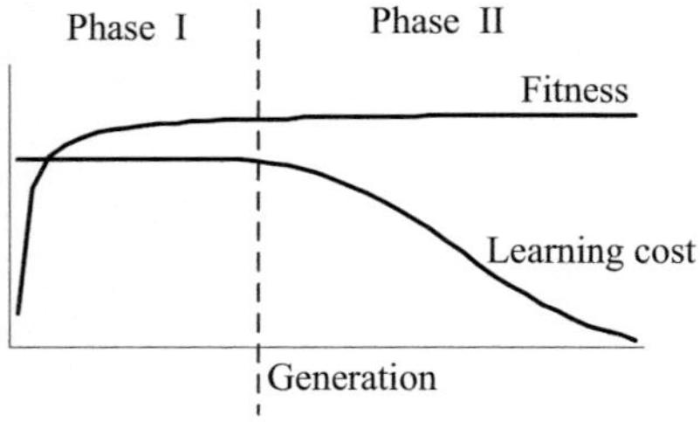


**Figure 1.12:** Two-phase process of the Baldwin effect.

the fitness of each individual in the population, followed by applying crossover and mutation to the surviving individuals in a repeated process of altering individuals. In case of Baldwinian evolution, individuals undergo learning before evaluating fitness, and the results of this learning are recorded as fitness. Additionally, in Lamarckian evolution, the learning outcomes are further inherited by newly generated individuals [54].

Applying Lamarckism to evolution leads to a faster rate of evolution by transferring the results of learning to the next generation. On the other hand, it narrows the solution search space and tends to lead to local solutions. On the other hand, applying the Baldwin effect avoids this problem and allows the next generation to make effective use of what it has learned. The choice depends on the nature of the problem. For simple problems, the Lamarckian approach is advantageous, while the Baldwin effect is advantageous for complex problems with many local solutions. For example, Geoffrey Hinton[5] tested the Baldwin effect using evolutionary computation [50]. They demonstrated the co-adaptation (i.e., smoothing of the fitness landscape) in the "needle in a haystack" problem.

## 1.2 Large Language Models (LLMs) and Generative AI

Generative AI is a technology that uses deep learning and machine learning methods to automatically generate human-like text, images, music, video, and other digital contents. Large Language Models (LLMs) are a type of generative AI that specializes in natural language processing, learning from vast amounts of text data to achieve a higher level of language understanding.

LLMs have increasingly shown behaviors resembling human language capabilities, demonstrating significant progress in the field of artificial intelligence. These models have been successfully applied in diverse tasks such as translation, coding, and question answering. The current landscape of LLMs is marked

---

[5]  Geoffrey Everest Hinton (1947–): A computer scientist born in England. Engaged in research on neural networks; known as the Godfather of AI. He won the Nobel Prize in Physics in 2024.

by a wide array of variants, each tailored to specific applications, languages, and purposes. These models vary not only in size, ranging from very large to comparatively tiny models, but also in their intended use cases, such as question-answering chatbots, coding assistants, legal analysis, and more. In addition to these variations, there have been significant augmentations to LLMs, enhancing their capabilities. These enhancements include the integration of code interpretation, which allows models to generate and compile programming codes, and multi-modality, where models can process and generate both text and images. Other notable advancements include the ability for models to utilize external tools, perform document retrieval, and more, broadening the scope of tasks LLMs can handle. Despite these advancements, there remains a challenge in effectively utilizing these generally pretrained LLMs in interactive environments that require advanced skills such as reasoning, planning, or decision-making. Reinforcement learning is a viable solution to such problems, but it requires a large amount of data and lacks interpretability. This gap calls for novel approaches to leverage the potential of LLMs beyond their current applications.

One of the blind spots of LLMs is that they are not good at arithmetics or elementary mathematics. LLMs are "probabilistic parrots" and only output a probabilistic mixture of what they have been told (data stored on the internet web) [9]. Therefore, LLMs can give correct answers to questions for which data is on the web, but they also make mistakes or give hallucinating answers to simultaneous equations and even simple algebraic calculations. In addition, LLMs are not good at logical thinking and are incapable of mathematical reasoning such as syllogisms, proof by contradiction, and inductive reasoning.

On the other hand, humans are good at mathematics and abstract thinking. Does this really come from the accumulation of experience? If not, then the expectation of mathematical achievement in LLMs may be misplaced.

In addition, the following shortcomings and questions also have been raised about LLMs:

- Correctness cannot be guaranteed. Since it is a statistical method, it does not always give the best answer.
- Development cost is high. For a network with billions to trillions of parameters, the language structure and context are learned using hundreds of TB of text data.
- Prompt engineering is required. This requires artisanal trial-and-error and empirical knowledge. Prompt engineering is difficult to study from a scientific or algorithmic point of view.
- There are biases in the data. Therefore, LLMs' answers may also be biased. The answers may be tainted by severe slander and hate.

- It is difficult to exchange information among different LLMs. It is not easy to integrate multiple LLMs. For example, in the past, Google Search Engine did not disclose information. This can hamper research and development.
- The ways LLMs answer differ from human thinking. In particular, there are differences in cognitive bias. (see Section 6.6 for details).

This book describes several ways to improve LLMs through the concepts of swarm and evolutionary intelligence. For example, in Section 5.4, we introduce prompt engineering by means of meta-heuristics. We will examine the possibility that LLMs and humans can cooperate with each other, leading to new ways of thinking in Section 3.7.

# Chapter 2

# Evolutionary Approach to AI

"A slow sort of country!" said the Queen. "Now, here, you see, it takes all the running you can do, to keep in the same place. If you want to get somewhere else, you must run at least twice as fast as that!" "I'd rather not try, please!" said Alice. "I'm quite content to stay here? Only I am so hot and thirsty!" (*Through the Looking-Glass, and What Alice Found There* by Lewis Carroll).

## 2.1 Co-evolution

One example of endless evolution is "co-evolution." This section describes co-evolution in the biological world and its computing method.

As a result of co-evolution, one of the following is established between organisms:

1. competition (both harming each other)
2. cooperation/symbiosis (both benefit from each other)
3. cooperation (only one benefits while the other does not)

Generally speaking, the transition from parasitism to cooperation occurs in the course of evolution. In other words, a host that initially excludes the parasite cooperates with the parasite at a certain point.

In co-evolution (especially competition), the "Red Queen hypothesis" has been proposed [111]. In "Alice in Wonderland," the world around the Red Queen is always moving. Therefore, to stand still in this world, one must be in constant motion. To run in this world, one must run twice as fast as in the normal world. The same is true in the evolution. Every species of organism is constantly evolving. In order to survive (stand still) in the biological world, it is necessary to keep evolving (keep running). If it stops evolving, the species will have no choice but to become extinct. Such a theory is called the Red Queen hypothesis. This phenomenon is not limited to the biological world. Examples can be found in the race to develop automobiles, the arms race among major powers, etc.

We will explain co-evolution with a real-life example. It is the relationship between the giant tortoise (Chelonoidis niger) and the prickly pear cactus in the Galapagos Islands. The giant tortoise feeds on the fruits and buds of the prickly pear cactus. When the giant tortoises eat the fruits, they expel the indigestible seeds in their feces. As a result, the seeds are dispersed far away as the tortoises move. This is convenient for the cactus. However, the loss of young shoots eaten by the giant tortoises is a major loss, and even in danger of extinction. Therefore, the cactus must somehow prevent the giant tortoise from eating its young shoots. The cactus's strategy was to grow taller and taller. However, the giant tortoise also stretches its neck and eats the buds that are about several tens of centimeters high. This is where the co-evolution (arms race) between the giant tortoise and the prickly pear cactus began. What happened as a result of co-evolution? The prickly pear cactus evolved the following strategy:

1. Grow high above the ground (can reach 3–4 m).

2. Wear thorns until they grow to about 2 m.

This made it impossible for the giant tortoises to eat cactus buds, and they were forced to change their diet. Why do cacti deliberately drop their thorns when they reach over 2 meters in height? This is probably because it is expensive to maintain the spines. When the thorns are no longer needed, the energy supply is cut off and the thorns fall off automatically. One evidence of such co-evolution is that on islands without giant tortoises, the cacti do not grow taller than about 1 meter, and their spines are soft.

Figure 2.1(a) shows the Galapagos giant tortoise. The tortoises have different carapaces on each island, resulting in separate lineages (subspecies). Figure 2.1(b) is a cactus from Santa Fe Island, which can be seen to be 2 to 3 meters tall. The lower part of the plant is covered with hard spines. The plant in the center of Fig. 2.1(c) is also a cactus, and its lower thorns have already fallen off, exposing the bark of the tree. On either side of this tree is the same kind of cactus brought

(a) Galapagos giant tortoise

(b) Prickly pear cactus

(c) Various cacti

(d) Cactus-eating tortoise (lonesome George)

**Figure 2.1:** Co-evolution in the Galapagos Islands.

from another island where no giant tortoises live. As you can see, there is a clear difference between these cacti. In Fig. 2.1(d), a giant tortoise stretches its neck out to eat a cactus. This tortoise, known as lonesome George, was the sole survivor of Pinta Island. Researchers were trying to keep the Pinta Island lineage alive, but George seemed to be unwilling to do so and preferred solitude.

Note that the results of co-evolution cannot be easily reversed. The giant tortoise is already extinct on Santa Fe Island, but the prickly pear cactus, having already evolved the above strategy, can no longer be made shorter.

This concept of co-evolution is extremely useful in the application of evolutionary computation. In the following, we will look at some examples of co-evolutionary methods for sorting networks.

Designing efficient sorting algorithms is a fundamental problem in computer science. The goal of sorting is to sort data elements according to some regular order (e.g., numerical or alphabetical) in a minimum amount of time. One algorithm for this is a sorting network. This is the sorting of a list with a certain finite number ($n$) of elements using parallel devices (i.e., comparators shown in Fig. 2.2(a)). When a comparator receives two inputs ($x_1,x_2$), it outputs the larger one (on the upper side in the figure) on one side, and the smaller one on the other (on the lower side). Figure 2.2(b) shows a sorting network for a list with $n = 4$

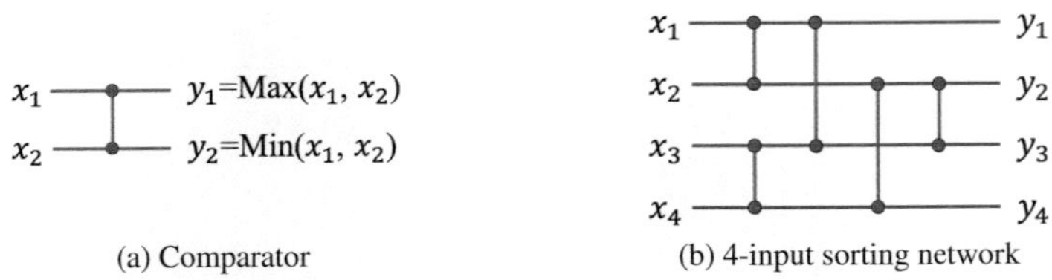


**Figure 2.2:** How to sort a list of numbers?

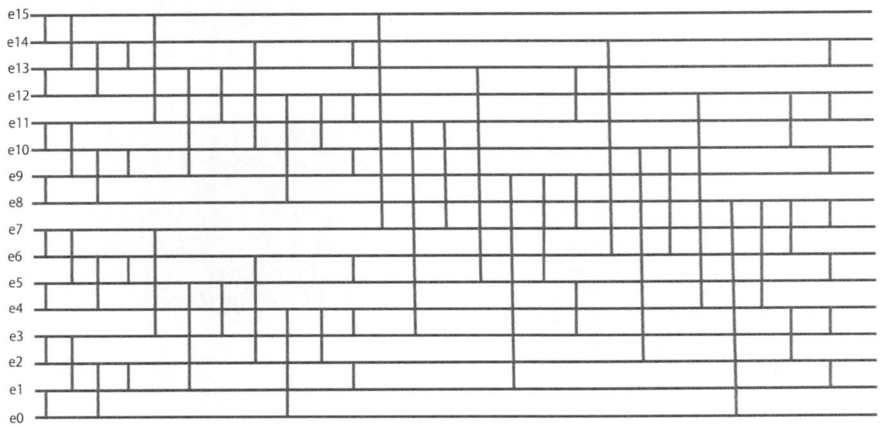


**Figure 2.3:** Batcher's sort.

elements. When this network receives 4 inputs (i.e., $x_1, x_2, x_3, x_4$) from the left, it gives a sorted output (i.e., $y_1, y_2, y_3, y_4$) from the right.

Figure 2.3 shows a sorting network for a list with $n = 16$ elements. This is called Batcher sorting. The horizontal lines in the figure represent elements of the list. The vertical segments are comparators made between two intervening elements. For example, the leftmost row of segments indicates comparisons made between e0 and e1, e2 and e3, etc. If the compared elements are not in the desired order, they are reordered. To sort the elements of the list, the data is moved from left to right in the network, making (and replacing, if necessary) the comparisons described vertically. The vertical comparisons are independent and can be done in parallel. The goal of the sorting network is to perform the correct sort efficiently (i.e., with the least number of comparisons). In this Batcher sort, $n \log_2 n - 1$ comparisons are required to sort $n$ elements.

What is the minimum number of comparisons required to correctly sort a list of $n$ elements? In the 1960s, there was a research race over the case of $n = 16$. In 1962, Bose and Nelson designed a sorting network that required 65 comparisons to sort a list with $n = 16$. In 1964, Batcher, Floyd, and Knuth designed a sorting network that required 63 comparisons. This network is shown in Fig. 2.3.

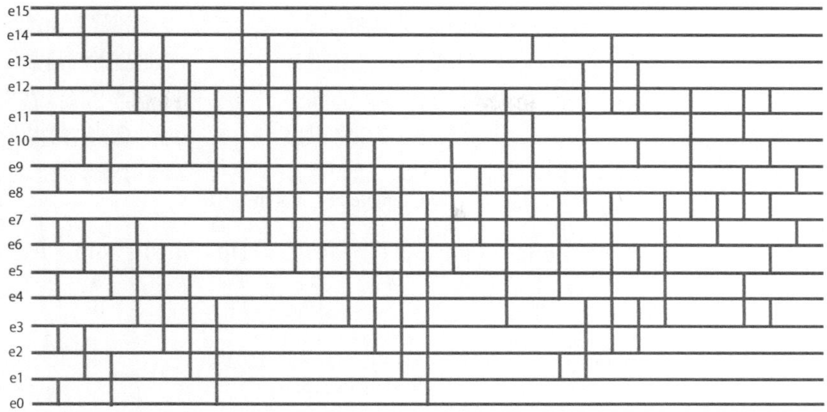


**Figure 2.4:** Green's sort.

**Table 2.1:** Best known sorting networks.

| Year | Inventor | #. of comparators |
|---|---|---|
| 1962 | Bose, R.C. Nelson, R.J. | 65 |
| 1964 | Batcher, K. Knuth, D.E. | 63 |
| 1969 | Shapiro, G. | 62 |
| 1969 | Green, M.W. | 60 |

Although this was considered minimal, in 1969, Shapiro presented a sorting network with 62 comparisons. In the same year, Green developed a network with only 60 comparisons (Fig. 2.4). The controversy died down after Green's discovery, although no proof of the optimality of these networks was given. Table 2.1 summarizes these results. Recent studies have proven that the lower limit for the number of comparators is 57 or more. Also, there is no known sorting network with fewer than 60 comparators.

Daniel Hillis [49] attempted to solve this problem using GA. First, he described the genotypes in the sorting network as pairs of ordered lists as follows:

```
(2,5),(4,2),(7,14),...........
```

These pairs represent a series of comparisons as follows:

1. First, compare line 2 and line 5, and if necessary, exchange them.

2. Then, compare line 4 and line 2, and, if necessary, exchange them.

3. ⋯

Note that the length of the genotypes is set from 60 to 200 (i.e., 60 to 200 comparators are considered).

| | | |
|---|---|---|
| 10110101011110011110010010101001<br>10110101001001110011110010101001 | 01001100101101001111010001100011<br>01010101011101001100011111001100 | 01100111100000001101001101000111<br>01101011011110011000000110100100 |
| 11010100111101010011110111011110<br>10111011010001010000000100010010 | 00011101001011110000101010110111<br>11000110100010100101111111111100 | 01100111101100110111100111001111<br>11100000101101011000011100101111 |
| 01011000101000110001110110111001<br>11111000111001001110010101001001 | 11011011001110101001001101010110<br>00110000110011001010111110000110 | 11010011011101001011000000010010<br>00011111101111100100110110111001 |
| 11010100111101010011110111011110<br>10111011010001010000000100010010 | 10111001100010010010101011011000<br>00010000011101011101010101100011 | 01010111001011011001110101101000<br>01010001000001100001101011111111 |
| 10100010110111100011101001010011<br>00010111111100001100000001010010 | 01100111100000001101001101000111<br>01101011011110011000000110100100 | 01011101001000000010111111111111<br>00101100000110100000000110010110 |

(a) Example of genotype

| | | | | | | | | |
|---|---|---|---|---|---|---|---|---|
| **codon** | 1011 | 0101 | 0111 | 1001 | 1110 | 0100 | 1010 | 1001 |
| | ⇓ | ⇓ | ⇓ | ⇓ | ⇓ | ⇓ | ⇓ | ⇓ |
| **integer** | 11 | 5 | 7 | 9 | 14 | 4 | 10 | 9 |
| | | | | ⇓ | | | | |
| **comparator** | | (11,5) | (7,9) | (14,4) | (10,9) | | | |

(b) How to express

| | | | | | | | | |
|---|---|---|---|---|---|---|---|---|
| **chromosome A** | 1011 | 0101 | 0111 | 1001 | 1110 | 0100 | 1010 | 1001 |
| **chromosome B** | 1011 | 0101 | 0010 | 0111 | 0011 | 1100 | 1010 | 1001 |
| **phenotype** | (11,5) | | (7,9)(2,7) | | (14,4)(3,12) | | (10,9) | |

(c) How to map from a genotype to a phenotype

**Figure 2.5:** Diploid genotypes.

Hillis used diploid genotype representation. Each individual consists of 15 pairs of 32-bit chromosomes (see Fig. 2.5(a)). Each chromosome consists of eight 4-bit codons. The codons represent integers 0 through 15, and the values indicate the positions of the 16 elements (i.e., lines) in the list (Fig. 2.5(b)). Adjacent pairs of codons in a chromosome describe a comparator between two elements. In other words, one chromosome encodes four comparators. As shown in the figure, one chromosome pair encodes from four to eight comparators because of duplication (Fig. 2.5(c)). Chromosome pairs are aligned and read from left to right. At each position, a pair of codons in chromosome A is compared to a pair of codons in chromosome B. If they encode the same number of pairs (i.e., homozygous), then only one pair is phenotypically expressed. If they encode different numbers of pairs (i.e., heterozygous), then both pairs are phenotypic. Thus, 15 pairs of chromosomes produce phenotypes that make a minimum of 60 and a maximum of 120 comparisons. More homozygous positions on a chromosome means that fewer comparisons appear in the sorting network. Remember that the goal is to find a minimum sorting network equal to the network by Green. Therefore, it is necessary for GA to find individuals who are homozygous at all positions in the correct sorting genotype. Note that this coding method cannot find a sorting network with less than 60 comparators.

In this experiment, the initial population consisted of individuals randomly generated from the genotypes. However, there are the following exceptions. Most of the known 16-element minimum sorting networks start with the same 32 comparison patterns. Thus, the first eight chromosome pairs were fixed to encode

**parent1 (diploid)**

```
A:01011101|00100000001011111111111     C:10111001100010010 01|0101011011000
B:00101100|00011010000000110010110     D:0001000001110101110|1010101100011
```

⇓

**gametes**

```
0101110100011010000000110010110     1011100110001001001101010110 0011
```

⇓

**child (diploid):**

```
0101110100011010000000110010110
1011100110001001001101010110 0011
```

**Figure 2.6:** Reproduction in diploid genotypes.

these comparisons. This is an example of using knowledge of the problem domain to facilitate the search for GAs.

Hillis defined fitness for a phenotype (i.e., a sorting network) as the percentage of correctly sorted elements in the list. Since there are so many input cases in the network, it is difficult to test all candidates. Therefore, each individual is tested by a set of randomly selected input examples in each generation.

Hillis used a simple GA with some modifications. The individuals are arranged in a two-dimensional grid. The spatial concept of distance between individuals is introduced. The purpose of placing the population on a spatial grid is to facilitate "seed formation" within the population. In other words, different types of networks are generated in different spatial locations, rather than all individuals converging on similar networks.

To create the next generation, half of the individuals with low fitness are eliminated and replaced by the surviving and replaced by the surviving higher-fitness neighbors. In other words, individuals with high fitness are given the opportunity to reproduce. Each individual is then paired with another spatially adjacent individual to produce offspring. Unlike in monoploid organisms, recombination in diploid organisms occurs between pairs of chromosomes (see Fig. 2.6). In a pair of 15 chromosomes, the crossover point is randomly selected. The gametes are generated by replacing the first codon of the first chromosome pair with the second codon of the second chromosome pair. As a result, the 15 gametes generated from the first parent are combined together with the 15 gametes from the second parent to produce one diploid offspring. This procedure is similar to sexual reproduction among diploid organisms as found in nature.

The mating described above is repeated as many times as necessary to form a new population. New individuals are stochastically mutated. This process is then repeated for several generations.

The above fitness is not necessarily sufficient to create the smallest (i.e., minimum) network. However, there should be an indirect selection pressure with respect to minimization. As in nature, homozygotes can protect important comparisons. If there are important comparisons at hetero positions in the chromosomes, they will be lost by crossover. On the other hand, if there are important comparisons at homologous positions, they will not be lost due to crossover. homozygous position, it will not be lost due to crossing over. For example, in Fig. 2.6, the leftmost comparison of chromosome B (the leftmost 8 bits encoding comparison (0,5)) is in a heterologous position and is lost in replication (i.e., the gametes have obtained the leftmost comparison of chromosome A). However, the rightmost comparison of chromosome A is in a homologous position. Thus, its chromosome is conserved even though the gametes obtain the rightmost comparison from chromosome B. In general, when an important comparison is found, it is very useful to place it in the homologous position. And as the homo position increases, the size of the resulting network decreases.

Hillis used a GA on very large populations with up to 1 million individuals. Runs were made up to 500 generations. The obtained smallest network evolved was the same as Bose and Nelson's network with 65 comparisons. This is 5 more comparisons than Green's network.

GA found a number of fairly good solutions (65 comparisons), but did not find any smaller networks. In other words, GA converged on local solutions. This is partly due to the insufficient test cases used to calculate the fitness values. GA discovers network strategies that work for the test cases, but the difficulty of the test cases does not change with each generation. Therefore, there is no selection pressure to change the local strategy.

To solve this problem, Hillis used the idea of co-evolution. The sorting network is considered the host, whereas the test case is considered the parasite. The idea is then to have an armed race between the two. The network population is modified to co-evolve on the same grid as the parasite population. The parasites consist of 10 to 20 test cases. Both populations also evolve under the GA. The fitness of the network is the percentage of correctly sorted parasite cases placed on the same grid position. The parasite fitness is given as the proportion of cases that baffled the network (i.e., the network failed to sort correctly).

The evolution of the test case population drives the evolution of the network population. The more correctly the network sorts, the more difficult the test cases become and the more specialized they evolve to exploit weaknesses in the network. This effect also promotes changes in the network population. In other words, instead of settling for the same local solution, the network is constantly changing in search of new strategies. This is the "Red Queen hypothesis." As a result, a sorting network with 61 comparisons was found. However, it fails to outperform the Green's network.

A recent study reports that Green's network was found by evolutionary computation [149]. This is based on the novelty search described in Section 2.6. The idea of co-evolution is similar to GANs in deep learning (see Section 2.2.2).

## 2.2 IEC for generative AI

Let us consider the application of EC to problems such as designing tables that match the atmosphere of a room or composing unobtrusive ringtones for mobile phones.

EC might seem applicable to these problems if they are considered in terms of optimizing the size and color of the boards in the case of the tables, or the frequencies, filters, and other parameters in the case of the synthesizers. However, the problem in this situation is how to evaluate each unit. An evolutionary system based on the survival of the fittest must include a method for evaluating whether individuals are suitable for their environment; in other words, how close they are to the optimal solution. For example, when using GA to evolve to a solution to the shortest path for the traveling salesman problem (TSP; see 96 page), the solution represented by each individual is assigned a degree of fitness in terms of path length. Is it possible to use a computer in the same manner to determine whether a table matches the atmosphere of a room? Unfortunately, modeling a subjective evaluation process based on human preferences and feelings, and then implementing such a model on a computer, is an extremely difficult task.

For instance, consider the fitness function in the case of a computer drawing a portrait. Conceivably, the first step is to compute the Euclidean distance between the structural information of a face extracted from a photograph and the portrait drawn by the computer. However, drawing a portrait is not the same as creating a photorealistic representation of a face–rather, drawing a portrait involves the more interesting process of capturing the unique features of the face and then drawing them in a stylized manner. But can a computer determine whether a person is identical in a facial photograph and a portrait in which specific facial features are captured and stylized? This fitness problem is considered to be extremely difficult to implement using the existing level of technology.

Nevertheless, there is something close to all of us that is capable of performing such evaluations instantaneously: the brain. Humans have evolved a method for direct evaluation of individual units, and in this regard, the human evaluation system can be incorporated into an optimization system as an evaluation function. Thus, when EC is employed for optimization based on human subjective evaluation, this process is then referred to as interactive evolutionary computation (IEC) [11].

### *2.2.1 Biomorph and IEC*

IEC is EC in which a human is substituted for the fitness function. In IEC, the person using the system (the user) directly evaluates each unit, and the viability (fitness) of subsequent generations is derived depending on the user preferences (Fig. 2.7). When implementing this system, personal preferences and sensations are not modeled, and evaluation based on user subjectivity is incorporated into

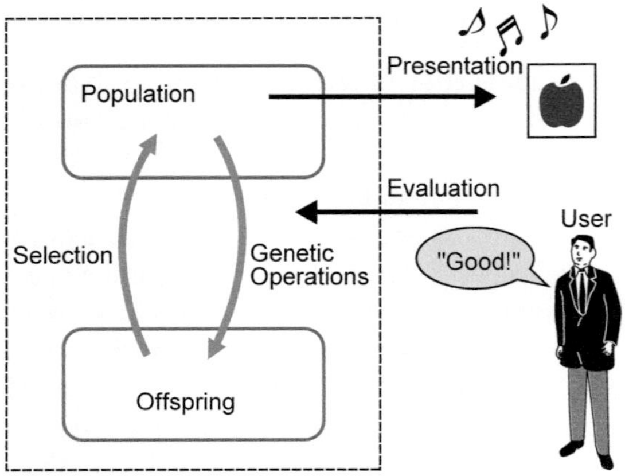


**Figure 2.7:** Overview of IEC.

the system as a black box. In contrast to conventional EC, where evolution is modeled as the result of a struggle for survival, IEC is inspired by the process of humans' breeding agricultural products and domestic animals.

Modern keywords such as "emotional (kansei) engineering" and "humanized technology" reveal the increasing tendency for technology to make use of human subjectivity. As part of this trend, IEC is beginning to attract attention as a method able to incorporate the user's subjective evaluation system.

Various types of human evaluation scales are used in IEC. For example, psychological experiments often use a scale with at most 7 grades, although a scale with 10 or 100 grades would certainly increase clarity in everyday situations. In contrast, for two-step evaluation, members of the population can be either selected (or not) as a parent for the next generation. This process is referred to as "simulated breeding" [148], since it is analogous to artificial breeding, in which desirable features are selected and then bred.

Let us see Biomorph (Fig. 2.8) as an example of IEC. This simulator is described in Dawkins' book Blind Watchmaker (see page 7 for details). Biomorphs can be regarded as the first example of IEC; specifically, evolution artificially created with a computer through selection based on human subjective preferences.

In this simulation, each generation consists of a group of males and females (12 insects each). The initial generation of insects is randomly generated. The user selects the parent that should give birth to the next generation from the male and female population, respectively. The large window on the right (Display Window) shows the state of the insects. The top 12 individuals ($3 \times 4$) are the male population and the bottom 12 individuals ($3 \times 4$) are the female population. Each insect gene (length $2n$) is a Fourier coefficient ($A_1 \cdots A_n B_1 \cdots B_n$) that controls the drawing of the two-dimensional X and Y coordinates. In this case, the insect form is drawn while moving $T$ according to the following equation:

$$X = \sum_{i=1}^{n} A_i \cos(i \times t), \tag{2.1}$$

$$Y = \sum_{i=1}^{n} B_i \sin(i \times t). \tag{2.2}$$

Where $n = 8$ in this program. The insects are line symmetric because of the periodicity of the trigonometric functions.

In this system, a male and a female mating produce one male offspring and one female offspring, respectively. The offspring's genes are the result of standard GA crossovers and mutations acting on the parent's genes.

For example, let us run the following criteria to select the parents.

1. Male parent is selected as round as possible.
2. The female parent is randomly selected with the same probability.

Continuing this process, after a few dozen generations, a population like Fig. 2.9 evolved. As a result, the shape of many male parents had rounded features.

As seen in this example, repeated artificial selection can produce the desired structure. Furthermore, through selection by female preference for males, it can be observed that the shape of the male changes rapidly. This phenomenon of accelerated evolution through female preference is sexual selection, which will described in Section 2.3.

IEC established a new method of creation by user choice. Biomorph has inspired many artists and researchers to apply it to the field of art, especially computer graphics (CG). The results are diverse, including portrait generation, graphic art, 3D CG generation for virtual reality spaces, animation creation, rhythm generation [144], music creation [5], and more.

As a typical example of graphic art based on IEC, we explain Sbart [148], i.e., a 2D image generation system using GP. Sbart performs rendering based on a 3-dimensional vector equation that includes the variables $x$ and $y$. By substituting the $x$ and $y$ coordinates of each point (pixel) in the image into the formula and converting the obtained values into drawing information as appropriate, a psychedelic and beautiful two-dimensional image is created (see Fig. 2.10). Specifically, each component of the vector obtained by substitution is painted in correspondence to Hue, Saturation and Brightness in the HSB color space. By introducing time as another parameter, it is possible to produce moving images that change over time.

### 2.2.2 DeepIE

IEC is used to help people make good designs [135]. For example, IEC has been applied to artistic fields such as dress design [65] and website design [94].

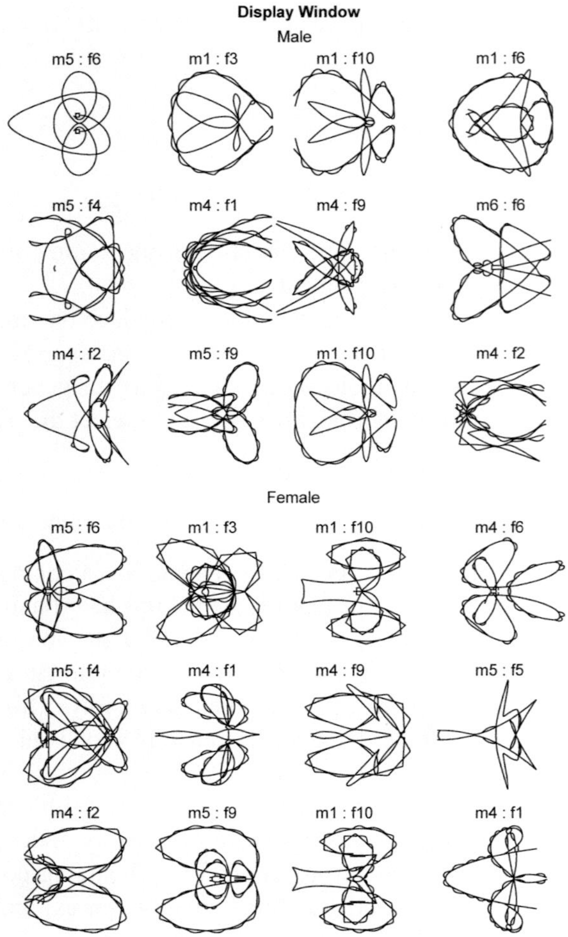


**Figure 2.8:** Biomorphs simulator (initial generation).

Some studies considered using GANs (Section 2.2.2) to combine deep learning with IEC. This approach is called "deep interactive evolution."

DeepIE [42, 135, 16] utilizes EC to solve the optimization problem by evaluating each individual using a fixed evaluation function to determine which genes to leave to the next generation. The IEC component uses subjective evaluation to determine the genes of the next generation, thereby allowing users to manipulate the evolutionary process and incorporate individual preferences and demands. DeepIE has been shown to be effective in generating images of shoes, faces, chairs, portraits, and 3D structures [16, 139, 63].

Let us consider GAN as related to images. At this time, the images that are exported per the latent variable $z$ value vary, but it is difficult to search for the latent variable space $z$ when targeting a specific image. This is caused by the la-

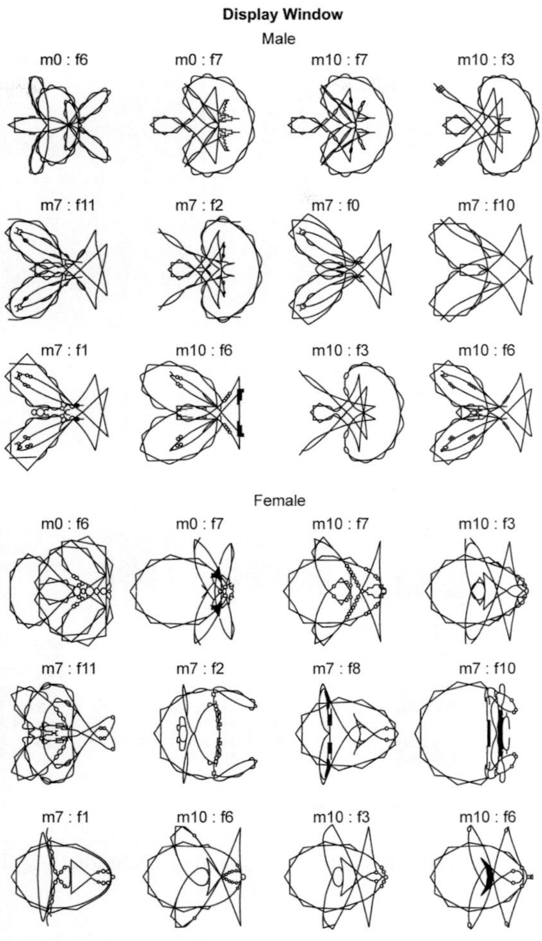


**Figure 2.9:** Biomorphs simulator (later generation).

tent variable $z$ being a vector that exceeds at least 10 dimensions (occasionally 20 dimensions). Moreover, the product traits that each dimension is acting upon are unclear. Therefore, the method of using evolutionary computation for searching for this latent variable space is being studied and is called "latent variable evolution." For example, CPPN2GAN [119] enables stage generation in games (i.e., Mario and Zelda) by combining CPPN and DCGAN. Additionally, deep master print [15] enables the generation of a master fingerprint (a fingerprint that matches any person's fingerprint and that can trick the fingerprint authentication system) by combining CMA-ES and GAN.

The deep interactive evolution framework [16] is shown in Fig. 2.11. The overall flow is given in **Algorithm 1**.

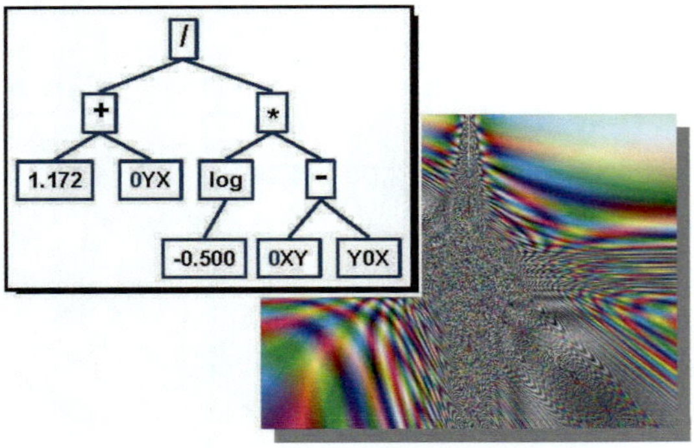


**Figure 2.10:** Sbart [148].

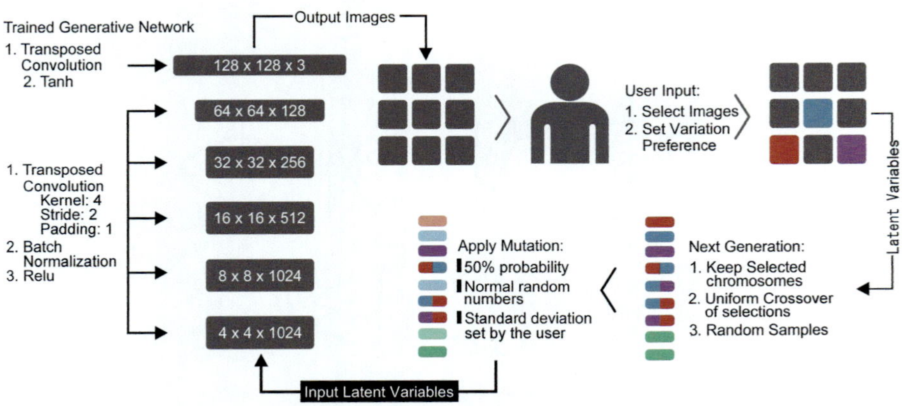


**Figure 2.11:** Overview of Deep Interactive Evolution (DeepIE) [16].

Deep interactive evolution uses general crossover, along with WGAN for a GAN model. Figure 2.12 illustrates examples of sneakers and face images evolved using deep interactive evolution [16]. Target images are shown in the left column. Final images are the last selected image. Best images are the images that the user selected as best from a compilation of all their selections over the entire evolutionary process.

## *2.2.3 DeepIE3D*

DeepIE3D [161] uses deep interactive evolution and WGAN-GP [44] to generate 3D models. WGAN-GP employs a gradient penalty term related to the critic's slope added to the WGAN loss function.

Interactive 3D modeling [81] is a framework of deep interactive evolution developed to facilitate users' 3D modeling. The difference with the traditional GAN structure is that, in addition to a generator, a phase called "projection" is

**Algorithm 1** Deep Interactive Evolution.

1: Create a generator by training a GAN.
2: Generate $m$ latent variables $z$ using the standard normal distribution. ▷ Generation of initial population.
3: **while** the user's preferred images are not generated **do**
4: Input latent variable $z$ to the generator and obtain an image corresponding to $z$.
5: Show the exported images to the user to select his/her preferred image(s). ▷ The user can choose multiple preferred images.
6: Perform genetic operations such as mutations and crossover. ▷ The latent variables of the selected images are considered to be genes with high fitness values.
7: The generated new genes are set as the latent variables at the next generation.
8: **end while**
9: Output the generated image from the obtained latent variable $z$.

added. This is similar to the voxels structure of input $x$ and provides the latent variable $P(x)$ that generates a realistic product. Specifically, a projection model $P(x)$ is designed to output latent variable $z$ while keeping in mind the following two points:

- Structure $G(z)$ generated from $z = P(x)$ should look similar to $x$. With this, edits matching the user's input are prioritized. For example, if the user selected an armchair when designing a chair, the model would prefer chairs with arms.
- Structure $G(z)$ should be feasible. This means that the generation of export $x' = G(P(x))$ is prioritized so that it does not differ from the GAN training examples.

The structure generated via the user stacking voxels is used as the input for interactive 3D modeling [81]. Instead of generating a structure from scratch, the main goal is to provide a rough skeleton and shape. The overall flow is described in **Algorithm 2.**

The focus in interactive 3D modeling is on the "realism" of the generated structure. Figure 2.13 shows a realistic structure $z$ and nonrealistic structure $z'$ generated in the latent variable space. Let us take a look at the diagram of the latent variable space in the figure. There is an area shown in red and another area in yellow. While the shape of the chair generated based on the latent variables in the red area is physically reasonable, the chair generated from the yellow area has legs that are cut off in the middle and a handrail that is broken off in the middle, resulting in a floating part. It is undesirable to generate physically unreasonable

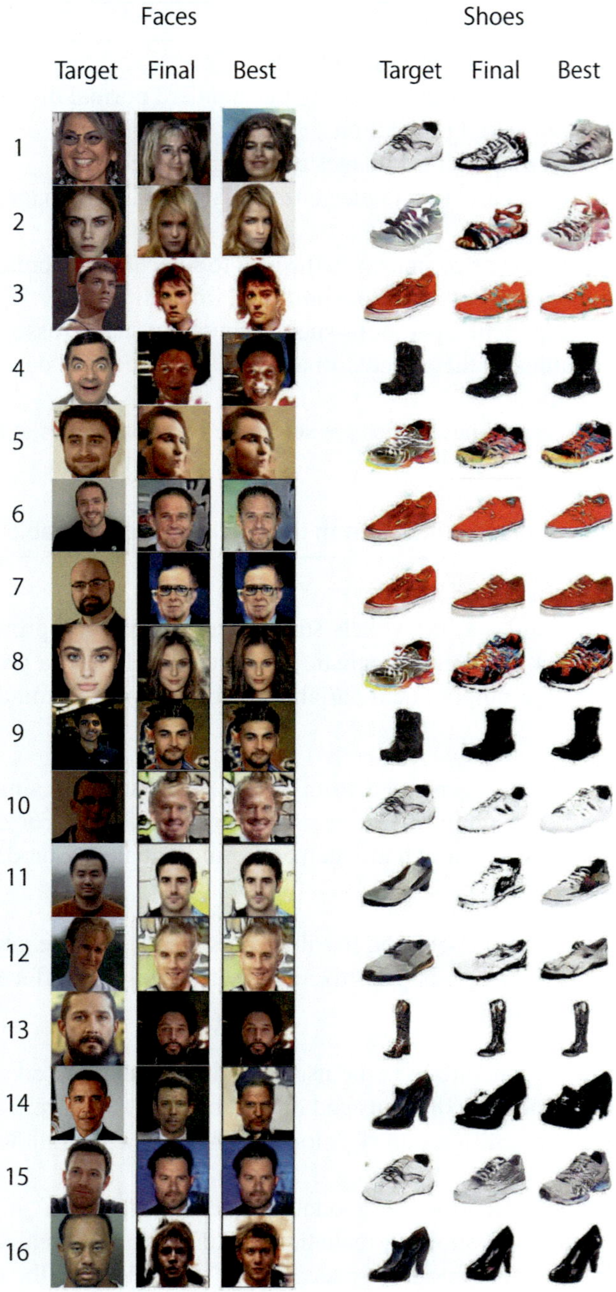


**Figure 2.12:** Evolved images by DeepIE [16].

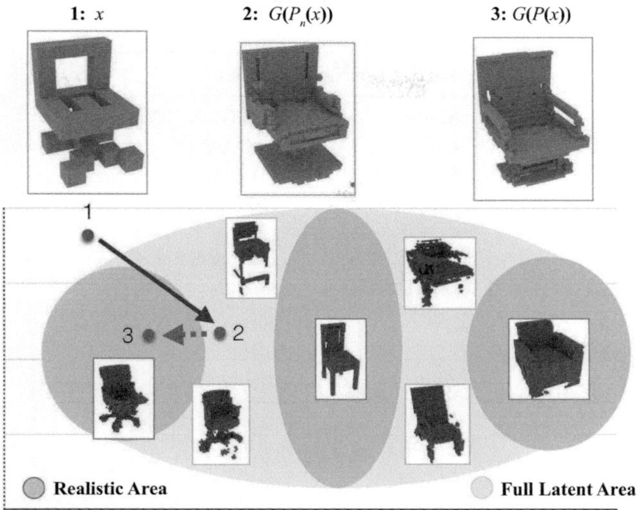


**Figure 2.13:** Depiction of how subcategories separate into realistic regions within the latent variable space of a generator [81].

---

**Algorithm 2** Interactive 3D modeling.

---

Generate generator: $G(z)$. ▷ GAN training.
Generate projector: $P(x)$. ▷ Create projection.
Set $t := 1$.
The user generates the structure $x_t$ by stacking voxels.
**while** the user's preferred image(s) is(are) not generated **do**
    Obtain $x_{t+1}$ from $x_t$ as follows:

$$z_{t+1} = P(x_t) \tag{2.3}$$

$$x_{t+1} = G(z_{t+1}) \tag{2.4}$$

    The user edits $x_{t+1}$.
    Set $t := t+1$.
**end while**
Output the obtained structure.

---

structures such as the ones in the yellow area when using GAN to design any structure.

By utilizing the training system of $P(x)$, the mapping to the latent variable space is restricted to the realistic domain $z$. Particularly, the projection is executed

**Table 2.2:** Parameters for DeepIE3D.

| **Parameter Name** | **Value** |
|---|---|
| Model | DCGAN [106] |
| Optimizer | ADAM [67] |
| #. of epochs | 10,000 |
| Batch size | 64 |
| #. of data | 129 |
| Dataset | shapenetcore v2[1] |
| Data format | voxel |

based on the following conditions:

$$P(x) = \underset{z}{\operatorname{argmin}}\ E(x, G(z)), \tag{2.5}$$

$$E(x, x') = \lambda_1 D(x, x') - \lambda_2 R(x'), \tag{2.6}$$

where $D(x_1, x_2)$ denotes the differences between two structures, $x_1$ and $x_2$. $R(x')$ is the evaluation function for the authenticity of the structure $x'$ generated by the generator. The output value of the GAN discriminator is used for this evaluation function (see [81] for details).

Let us examine whether there is a relationship between the discriminator output value and the realism of the structure when setting the product of DeepIE3D as the input for the discriminator. The main parameters we used for DeepIE3D are shown in Table 2.2. The results of generated chairs are shown in Fig. 2.14. Each of the discriminator's output values is in Table 2.3 (the figure and table correspond to the same $3 \times 3$ individuals). As can be seen from the results, Individual #1's value is 0.7319, which is outstanding large compared to the others. Individual #2 and Individual #4 have a value of a little more than 0.5. The other five individuals all have values slightly above 0.1. Let us recall that the closer to 1.0 the discriminator output values are, the more realistic. The closer to 0.0 they are, the more false they are determined to be; as a result, Individuals #1, #2, and #4 are determined to be more realistic. When observing the geometry of the exported structures, Individuals #1, #2, and #4 are chairs with a wide base and arms attached, while the other chairs have no arms and are narrow. However, from the point of view of realism, the Individual #1, #2, and #4 chairs all have legs that are cut off and appear unrealistic in some respects. The other five chairs do not have legs that are cut off for the most part and may actually be superbly realistic. Based on this, the discriminator does not determine the realism of chairs but shows whether it is similar to a chair with certain characteristics (here, a wide chair with arms).

[1]Dataset of 3D models with manually verified categories and alignment annotations. See `https://shapenet.org/`.

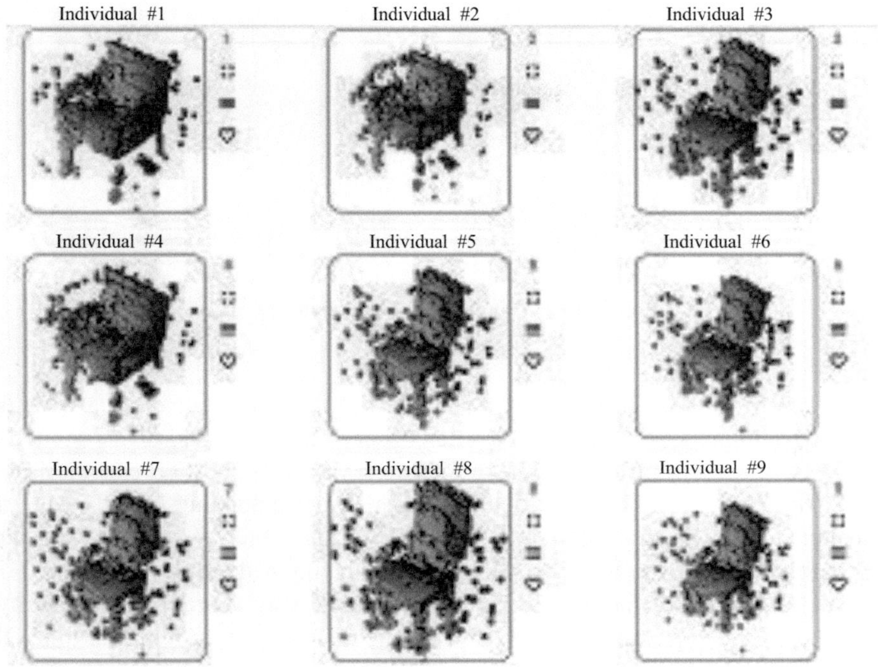


**Figure 2.14:** Individuals of generated chair structures.

**Table 2.3:** Results of applying the structures to a discriminator.

| | | |
|---|---|---|
| 0.7319 | 0.5156 | 0.1328 |
| 0.5249 | 0.1296 | 0.1506 |
| 0.1192 | 0.1236 | 0.1268 |

## 2.3 Sexual selection

Biologists are divided in their opinion regarding the excessive beauty of certain animals, e.g., peacock feathers, moose antlers, and zebra stripes, etc. Does it fulfill a function of distinguishing individual, of promoting mutual grooming, of protecting the body from harmful insects, or of adjusting the body temperature? Perhaps the aesthetics are a handicap?

Excessive beauty is not only limited to an animal's physical features and patterns. It can also be seen in the artistic skills displayed by animals. For examples, the Austrian blue-winged bowerbird (Ptilonorhynchus violaceus) attracts the attention of females by creating an elaborate structure called a "bower."

Even some fish display artistic expressions. In 2011, a new species of blowfish (Torquigener albomaculosus) was found in Amami isl., Japan (called "Galapagos of the East"). Although the length of a male specimen is just 15*cm*, the

**Figure 2.15:** A mystery circle (Amami isl., Japan).

fish constructs a complex radial-shaped nest at the bottom of the sea with $2m$-diameter tunnels (see Fig. 2.15). After observing such nests, divers started to refer to them as "mystery circles." In this mystery circle, two concentric circles with uneven surfaces are doubled, with stripes extending radially from the center. The blowfish scatter small pieces of shells, which they break with their mouths, into the central groove. When an egg-laying bed (circle) is satisfactorily completed, a female accepts to lay eggs in the nest for reproduction. The mystery circles are spawning nests (spawning beds) to appeal to females. Since the spawning of female blowfish depends on the quality of the nest, the artistic skill of the male is essential for the offspring to survive.

Researchers conceive that such wonderful patterns or artistic expressions might be unsuitable for adaptation. The maintenance of beauty requires energy (cost) and prevents the animal from hiding effectively when discovered by predators. The feathers of a peacock do not provide an advantage when running away from a threat.

Even Darwin was troubled by this problem when building the arguments for his theory of evolution. The following comment is know in a letter to his friend Asa Gray, a botanist at Harvard University, on April, 3, 1860 (a year after the publication of "Origin of Species"):

> The sight of a feather in a peacock's tail, whenever I gaze at it, makes me sick!

In this chapter, we preset several hypotheses explaining this, together with verification experiments based on genetic algorithms (GAs).

### 2.3.1 Sexual selection in relation to markers, handicaps, and parasites

In the following paragraphs, let us consider the difference in opinion between Darwin and Wallace in regard to the striped pattern of the zebra. First, let us look at Russel Wallace's statement.[2]

> The zebra is conspicuously striped, and stripes on the open plains of South Africa cannot afford any protection (an omission). Here we have no evidence of sexual selection, as throughout the whole group of the Equidae the sexes are identical in colour. Nevertheless he who attributes the white and dark vertical stripes on the flanks of various antelopes to sexual selection, will probably extend the same view to the Royal Tiger and beautiful Zebra [26, p.302].
>
> It may be thought that such extremely conspicuous markings as those of the zebra would be a great danger in a country abounding with lions, leopards, and other beasts of prey; but it is not so. Zebras usually go in bands, and are so swift and wary that they are in little danger during the day. It is in the evening, or on moonlight nights, when they go to drink, that they are chiefly exposed to attack; and Mr. Francis Galton, who has studied these animals in their native haunts, assures me, that in twilight they are not at all conspicuous, the stripes of white and black so merging together into a gray tint that it is very difficult to see them at a little distance [155, p.220].

In *On the Origin of Species*, Darwin argues that a female chooses a mate after seeing a display of marvelous plumage and unusual, outlandish body movements. Being chosen by the opposite sex in this way is known as "elimination through sexual selection" or simply "sexual selection."

> The law of battle for the possession of the female appears to prevail throughout the whole great class of mammals. Most naturalists will admit that the greater size, strength, courage, and pugnacity of the male, his special weapons of offence, as well as his special means of defence, have been acquired or modified through that form of selection which I have called sexual. This does not depend on any superiority in the general struggle for life, but on certain individuals of one sex, generally the male, being successful in conquering

2 
Alfred Russel Wallace (1823–1913): British naturalist, biologist, and explorer. He identified the Wallace line, a distributional boundary that divides the distribution of animals in Indonesia into two distinct regions. He and Charles Darwin independently and almost simultaneously devised the theory of evolution by natural selection, and the paper sent by Wallace from the Malay Archipelago surprised the slow writer Darwin. Eventually, both papers were presented simultaneously in 1858 at the Linnean Society after Darwin had made "a delicate arrangement" in relevant parts of his paper [13].

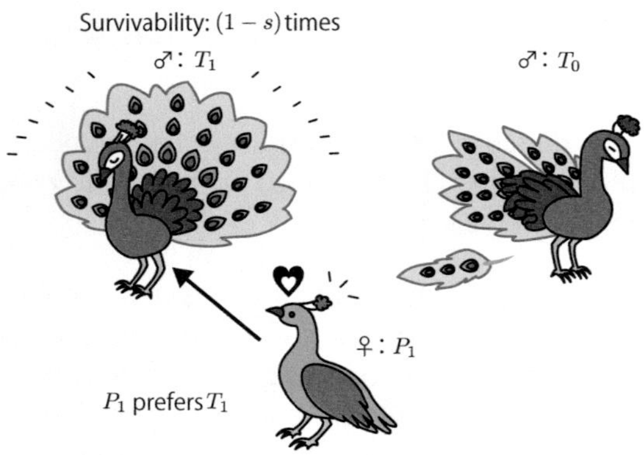


**Figure 2.16:** The sexual selection.

> other males, and leaving a larger number of offspring to inherit their superiority than do the less successful males [26, Ch.XVIII].

Ronald Fisher[3] supported the theory of sexual selection proposed by Darwin (Fig. 2.16). In a seminal work entitled *The Genetical Theory of Natural Selection*, Fisher presented an adequate explanation of the female animal preferences described by Darwin in his theory. The premise of Fisher's theory is that successful achievement generates further successful achievement, resulting in positive feedback with the potential for explosive growth. For example, if the preference for a long tail becomes increasingly successful, the tails of males in later generations will become longer, and females will prefer the males with even longer tails. This results in the development of yet longer tails and a preference for long tails in females. Since success depends on the rate of occurrence, the process progresses in a self-reinforcing manner.

The optimal strategy is that practiced by the majority, and mimicking the actions adopted by most individuals is thus preferable. Therefore, selection based on a long tail and the preference for a long tail progress together. The appearance of males and the preference in females for a certain appearance evolve hand-in-hand in such a way that they reinforce each other. This process is considered to have eventually given rise to the extraordinary tail feathers of the peacock, and explains why the evolution and preference of a certain appearance often exceeds a moderate level and becomes a runaway process [25, p.284]. Such rapid growth is referred to as "the runaway effect" by Fisher.

In contrast, Wallace adopted the point of view that coloration is ultimately derived from "protective colors and marks." This idea about protective colors explains both concealing coloration (which hides the body by deceiving the eyes

---

3  Sir Ronald Aylmer Fisher (1890–1962): One of the founders of population genetics as well as statistics, and was a geneticist and evolutionary biologist. Unfortunately, he was also an advocate of eugenics. His pioneering work in comparative and animal psychology is closely related to current AI research.

of an adversary) and conspicuous coloration. In other words, Wallace points out that although it might appear at first sight that the coloration of some animals is extremely noticeable, this coloration is in fact concealing in the natural habitat of the animal. Examples supporting this view include the zebra, the tiger, and the giraffe, which blend seamlessly into their respective native environments. In addition, inedible animals that use bright coloration as a warning, and animals that mimic such coloration, are in excellent agreement with this explanation. Furthermore, colored markings distinguish individuals of the same species, which serves as an aid for social animals (such as ants and bees) to stay together or to recognize potential partners. The bright coloration of various birds can also be explained in this way [25, pp.176-178].

Although all the details about coloration as proposed by Wallace do not necessarily hold true, and despite the large number of demonstrated counterexamples, the general outline of Wallace's argument is well constructed, and has been adopted as a fundamental viewpoint in Darwinian theory.

Conversely, the "handicap principle" proposed by the Israeli biologist Amotz Zahavi[4] accounts for the seemingly paradoxical existence of animals that pay an extraordinarily high cost for their appearance. With its extravagant feathers, the peacock makes a statement that it can afford its beautiful appearance, since its nourishment is plentiful and its agility and stamina eliminate the risk being discovered by predators. Thus, the peacock is in a superior position even with the disadvantage imposed by this handicap (Fig. 2.17). In the same manner, bucks with extraordinary horns make a statement that despite their handicap, their nourishment is sufficient and they are strong enough to win a fight.

> Why does the gazelle reveal itself to a predator that might not otherwise spot it? Why does it waste time and energy jumping up and down (stotting) instead of running away as fast as it can? The gazelle is signaling to the predator that it has seen it; by "wasting" time and by jumping high in the air rather than bounding away, it demonstrates in a reliable way that it is able to outrun the wolf [175, pp.xiii–xiv].

A third theory that emphasizes the role of parasites was recently proposed, sparking controversy. This theory was developed by the animal behaviorist William Donald Hamilton,[5] and resembles the handicap principle. Using the male's feathers as a clue, a female assumes that she is choosing a partner with superior genes. A male's genes determine its immunological resistance to parasitic

4  Amotz Zahavi (1928–2017): Israeli evolutionary biologist. He proposed the "handicap principle" in 1977.

5  William Donald Hamilton(1936–2000): English evolutionary and theoretical biologist. He proposed the theory of kin selection based on inclusive fitness and elucidated the biology of altruistic behavior.

**Figure 2.17:** The handicap principle.

**Figure 2.18:** The parasite hypothesis.

agents (bacteria, viruses and parasites), since maintaining large beautiful feathers is difficult if the body has been severely damaged by parasites. Therefore, choosing a male with spectacular feathers equals choosing that with strong resistance to parasites (and hence strong capability; Fig. 2.18). Although, up to this point, Hamilton's theory shares the same line of thought as the handicap theory, it differs drastically from the latter in the following sense. While predators in Zahavi's theory are more or less known, new parasites constantly emerge, and the parasites themselves are extremely diverse. Therefore, a considerably stronger resilience must be maintained than that necessary for avoiding predators. The magnificence of a male's feathers clearly shows its resilience against current parasites, and the female wishes to obtain this information by all means possible. Hence, the feathers of the peacock have acquired this extravagance, which also appeals to the female.

To verify his theory, Hamilton investigated several species of birds (Passeriformes living in North America) in terms of their resistance to parasites, by ranking on a six-level scale the plumage and the song complexity of males. The results were subsequently compared with research data on the concentration of trematodes and protozoans in the blood of these birds, and a correlation with the assigned ranks was investigated. Unsurprisingly, species with more extravagant feathers and more complex songs also had greater resilience to parasites. Investigating the feathers of the females, the correlation was almost indistinguishable from that observed in males. Hamilton's parasite hypothesis was thus considered to be verified. However, a number of problems with this experiment have been pointed out, and the debate continues.

We now turn our attention to the use of GAs in sexual selection research.

### 2.3.2 The Kirkpatrick model

Donald Kirkpatrick [69] studied a mathematical model of sexual selection based on population genetics In this model, each individual carries two types of genes denoted $T$ (traits, features) and $P$ (preferences), and each gene has a pair of alleles ($T_0$, $T_1$ and $P_0$, $P_1$, respectively). $P$ encodes the preference of a female with respect to trait $T$ of a male such that females carrying $P_i$ prefer males carrying $T_i$ ($i = 0, 1$). These preferences depend on the rate of occurrence, and the probability of a $P_i$ female breeding with a $T_j$ male is set as

$$P(T_0 \mid P_0) = \frac{a_0 t_0'}{t_1' + a_0 t_0'}, \tag{2.7}$$

$$P(T_1 \mid P_1) = \frac{a_1 t_1'}{t_0' + a_1 t_1'}, \tag{2.8}$$

$$P(T_1 \mid P_0) = 1 - P(T_0 \mid P_0), \tag{2.9}$$

$$P(T_0 \mid P_1) = 1 - P(T_1 \mid P_1). \tag{2.10}$$

Here, $t_k'$ denotes the rate of occurrence of mature $T_k$ males.

Let us assume that trait $T_1$ is detrimental to survival. This precisely reflects the scenario that predators can easily discover peacocks with beautiful feathers, which are preferred by peahens. We therefore consider it more difficult for males carrying $T_1$ to survive than males carrying $T_0$. Specifically, we take the probability of survival of $T_1$ males as $1 - s$ times that for $T_0$ males (where $0 < s < 1$ is a real number). Although males also carry $P$, this gene is not expressed in their appearance, and similarly for $T$ in females.

In the Kirkpatrick model, the initial population has the same number of males and females. In each successive generation, a proportion of $T_1$ males die before procreating; explicitly, $s$ times the number of $T_1$ males are killed. Each female then chooses a male for mating, where $P_i$ females prefer $T_i$ males following eqs. (2.7) and (2.8). After breeding, the next generation is produced, and this process is repeated for a certain number of generations.

Kirkpatrick performed a simulation without devising an actual version of this system. As a result, Kirkpatrick derived an equation where the rate of occurrence of $P_1$ females and $T_1$ males is balanced (a state of equilibrium, where no further changes occur). Here, taking $p_1$ and $t_1$ as the respective occurrence rates of $P_1$ and $T_1$, Kirkpatrick presented the following equilibrium equation.

$$t_1 = \begin{cases} 0 & p_1 \leq V_1, \\ \frac{(a_0 a_1 - 1)(1-s)}{(a_0+s-1)[a_1(1-s)-1]} p_1 - \frac{1}{a_1(1-s)-1} & V_1 < p_1 < V_2, \\ 1 & V_2 \leq p_1. \end{cases} \tag{2.11}$$

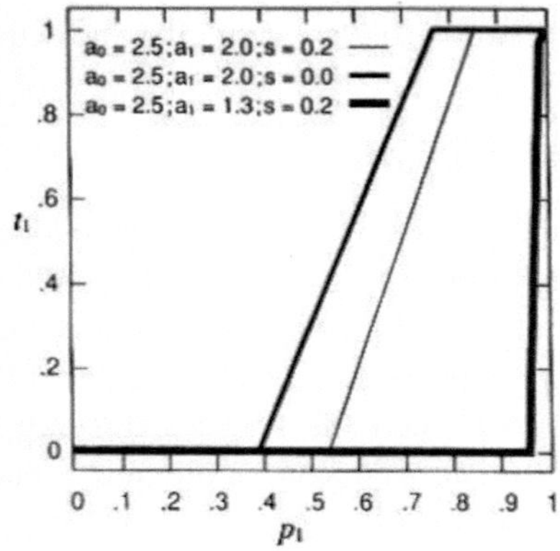


**Figure 2.19:** Kirkpatrick model and its equilibrium state.

Here,

$$V_1 = \frac{a_0 + s - 1}{(a_0 a_1 - 1)(1 - s)},$$
$$V_2 = \frac{a_1(a_0 + s - 1)}{(a_0 a_1 - 1)}.$$

In Fig. 2.19, eq. (2.11) is plotted with respect to several parameters. If $t_1$ and $p_1$ are now taken to be the respective proportions of $T_1$ and $P_1$ genes, depending on the parameters and the initial conditions, the occurrence rate of $T_1$ takes an arbitrary value between 0 and 1. For both $T_0$ and $T_1$ to be preserved in the population, the following condition must be satisfied:

$$1 - a_0 < s < 1 - \frac{1}{a_1}, \tag{2.12}$$

otherwise the population becomes stationary with respect to either $T_0$ or $T_1$.

The results indicate that an infinite number of equilibrium states exist for the occurrence rates of the genes. Hence, the equilibrium states lie on a curve rather than discrete points. Intuitively, natural selection should favor males carrying $T_0$; however, if the occurrence of $P_1$ is sufficiently high, $T_1$ males are preferred due to sexual selection, resulting in competition between natural and sexual selection.

In this model, changes in the occurrence rate of $P$ depend only on the occurrence rate of $T$, which shows that once the population has entered an equilibrium state, it remains on this equilibrium path. However, this situation does not necessarily hold true when the occurrence rates of the genes are altered by external forces (such as genetic drift, migration, or sudden mutation).

### 2.3.3 Simulation using GA

Collins and Jefferson simulated the Kirkpatrick model by using genetic algorithms [21], whereby each individual contained two chromosomes–one carrying gene $P$ and the other carrying gene $T$. In females, only $P$ was expressed, while in males only $T$ was expressed, and each gene took one of two values ($T_0$ or $T_1$ for $T$ and $P_0$ or $P_1$ for $P$). The population contained 131,072 individuals, with the same number of males and females at initiation. For each new generation, a fixed proportion of the $T_1$ males died before procreating, after which each female chose a mate from the remaining males, where mating consisted of gene recombination. Additionally, a sudden mutation could occur in all genes with a probability of 0.00001. This simulation weakened several assumptions of the Kirkpatrick model by assuming sudden mutations and a finite population size.

Figure 2.20 shows the results after 500 generations, where $a_0 = 2.0$, $a_1 = 3.0$, $s = 0.2$, $\mu = 0.00001$ and $N = 131,072$ are taken as the parameters values. The gene occurrence (of $T_1$ corresponding to $P_1$) in the final population is plotted over 51 iterations for various initial occurrences, and the results of Kirkpatrick's analysis are also included. The figure clearly shows that even with weaker assumptions the numerical results almost completely coincide with the analytical results.

Next, let us consider the evolutionary course until the population reaches equilibrium in the case where a new gene is inserted. Figure 2.21 shows the transitions for a population with an abundance of $P_1$ genes, but no $T_1$ genes. The initial generation thus had $P_1$ and $T_1$ proportions of $p_1 = 0.7$ and $t_1 = 0.0$, respectively, and the parameter values were $a_0 = 2.5, a_1 = 2.0, s = 0.2, \mu = 0.00001$ and $N = 131,072$. The marks $\diamondsuit$ in the figure indicate the states for every 50 generations. Although 100 generations were required for a sufficient number of $T_1$ genes to emerge, after that point the population rapidly converged toward equilibrium.

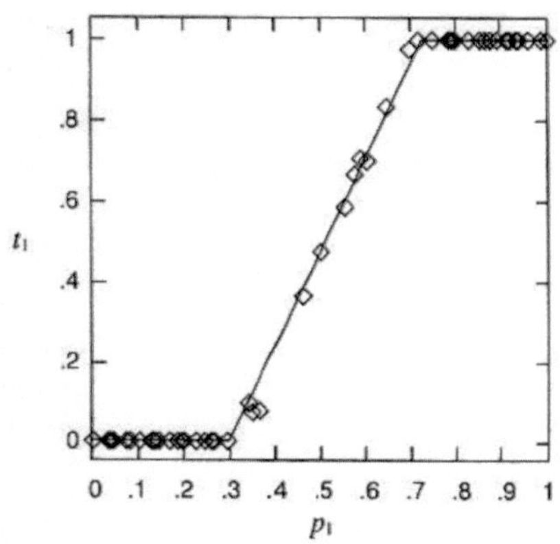


**Figure 2.20:** Evolutionary results after 500 generations.

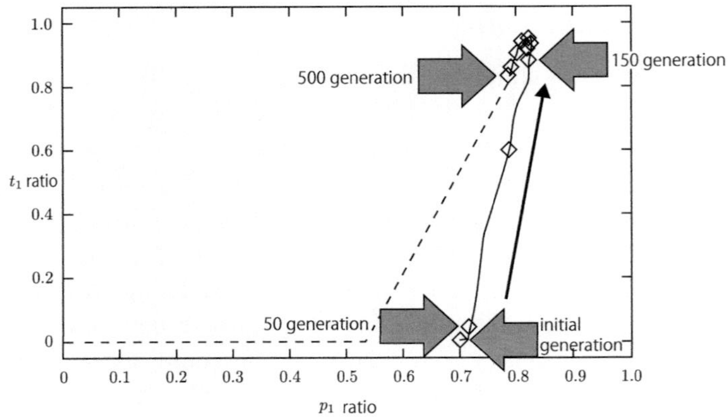


**Figure 2.21:** Evolutionary transition in the case where a new gene is inserted.

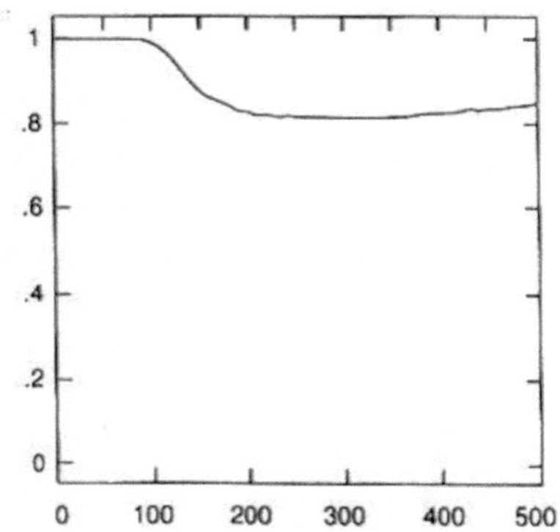


**Figure 2.22:** Survival ratios (male) vs. generations.

This experiment is a remarkable demonstration of the power of female preferences and sexual selection. The females are less interested in the males whose survival is easier ($a_0 > a_1$), and since these $T_0$ males are not chosen as often, eventually the number of males carrying $T_1$, whose survival is more difficult, increases rapidly and dominates the population. Figure 2.22 shows that the average number of surviving $T_0$ males decreases as the number of $T_1$ males increases. This result is a clear demonstration of the "runaway effect" proposed by Fisher. Regardless of the fact that $P$ and $T$ are located in different chromosomes, a positive correlation is formed between $P_1$ and $T_1$. The increased preference for $T_1$ is also shown by the simulation, with the proportion of $P_1$ increasing from 0.7 to over 0.8.

Moreover, Collins and Jefferson conducted experiments with weakened assumptions. In one of these experiments a spatial restriction for choosing a mating partner was imposed (a more realistic model, since most animals do not mate with distant partners). In another experiment, diploidic organisms were provided with dominant and recessive genes. These two experiments are difficult to understand from a mathematical viewpoint, and in both experiments Collins and Jefferson found considerably different behavior compared with Kirkpatrick's original model. In the case where a spatial restriction was imposed on mating, $T_1$ spread slowly across the population; however, in the case of diploidic organisms, $T_1$ was unable to spread across the population unless the gene was prevalent from the beginning.

Apart from the research mentioned above, several other intriguing studies on sexual selection have been reported. For example, Werner examined the changing of characteristics in a population as a result of sexual selection, claiming that these changes are almost unlimited [159]. In addition, Werner and Todd showed that sexual selection in which detrimental features are preferred can result in the extinction of a species, and he also conducted simulations where evolution of the signal used by males to attract females was modeled through sexual selection [160]. Geoffrey Miller[6] compared sexual selection to a venture company, and stated a theory emphasizing the possibility for progressive evolution that supports slow and conservative natural selection [86]. Furthermore, the application of sexual selection to engineering is also conceivable, for example, in interactive design systems, where certain types of runaway effect might prove to be productive. Such applications constitute an important topic for future research.

Enquist et al. [34] examined the effects of sexual selection using a simple neural network and evolutionary computation. In this experiment, they tested sex selection on the hen's cognitive system (i.e., neural network) and the length of the male bird's tail. The neural network consists of an input layer with a retina consisting of 6×6 receptive cells, a hidden layer with 10 nodes, and an output cell with one node (Fig. 2.23). If the output is above a threshold, the female has exhibited courtship behavior. Here, the weights of the network are mutated and evolved. There are different species of males that appear identical in all respects except for tail length. Suppose that a male of the same species as the female has a longer tail than a male of a different species. The female's cognitive system (neural network) is trained to distinguish long-tailed males (homologous) from short-tailed males (heterologous) and random images. The image is rotated to multiple positions and given to the input layer for learning. As a result, a network with almost no misrecognition has evolved. Table 2.4 shows the training and test image data. The network responds strongly to birds with longer tails (a, b) and even longer wings (c) in the testing case. However, it does not respond to the image without a tail (d). Even rather abstract patterns (e) are responded to more than normally. Thus, the network shows adequate generalization ability.

6  Geoffrey F. Miller (1965–): American evolutionary psychologist at the University of New Mexico. He is well known for his research on sexual selection and won the Ig Nobel Prize in Economics for his research that lap dancers who ovulate earn more tips. However, these studies are often criticized as being male-centered 45.

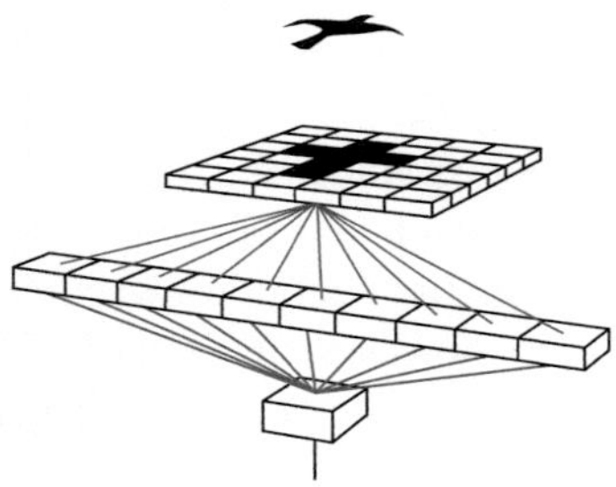

**Figure 2.23:** Cognitive system of a hen.

**Table 2.4:** Training and test data.

| Trainig data | | Test data | | | | |
|---|---|---|---|---|---|---|
| negative | positive | (a) | (b) | (c) | (d) | (e) |
| randam pictures | 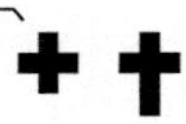 |  |  |  |  |  |

We now introduce sexual selection to co-evolve male tail length and female cognitive systems (see Table 2.5). In sexual selection, males are more compatible with those that induce strong courtship behavior in females, and females are more compatible with those that correctly recognize their own species. Males with longer tails are less likely to survive (more likely to die out) ("survival rate" in the table). Columns (a) to (e) show the results of sexual selection with different male survival rates. For example, in condition (a), males with tail length 4 are not allowed to survive. The experiment shows that, even though it is unfavorable for survival, males have longer tails and females are less likely to court males of the original species (see the "recognition rate" of the prototype in the table). This result indicates that the female's preference for the male's exaggerated trait (long tail) evolves out of the female's need to recognize the male's species.

Enquist et al. have experimentally verified that a similar mechanism of sexual selection leads to the evolution of an aesthetic preference for symmetry in animals [35]. Neural networks capable of recognizing rotated objects are optimally excited by patterns of radial symmetry. They concluded that star shapes, sun shapes, and eyeball patterns are preferred as sexual ornamentation traits. However, there have been criticisms of these studies to the effect that they do not actually learn geometric features and simply conform to the effect of averages [6].

In addition to the researches described so far, many other interesting studies on sexual selection have been published. Some species are known to reverse the role of females and males. For example, in Phalaropus lobatus or Banded pipefish

**Table 2.5:** Results of sexual selection.

| | Tail length | (a) | (b) | (c) | (d) | (e) |
|---|---|---|---|---|---|---|
| Survival ratios | 2<br>3<br>4 | 0.50<br>0.25<br>0 | 0.50<br>0.30<br>0.10 | 0.50<br>0.35<br>0.20 | 0.50<br>0.40<br>0.30 | 0.50<br>0.45<br>0.40 |
| Evolved phenotypes | |  |  |  |  |  |
| Recognition rate (tail length = 2) | | 100% | 88% | 84% | 58% | 59% |

**Figure 2.24:** A relative of the seahorse, the Banded pipefish (Syngnathidae).

(a relative of the seahorse (see Fig. 2.24)), the female engages in courtship behavior toward the male, while the male engages in egg-laying and brood-rearing. As predicted by the theory of sexual selection, the body color of the female is more vivid, while the male is smaller and more shabby.[7]

## 2.4 Adaptationism or evolving "beauty?"

In the previous section, we described the mathematical theory regarding sexual selection. We also discussed several hypotheses that are at odds with sexual selection. These fall into two main categories.

**Adaptationism:** Richard Dawkins, Alfred Russel Wallace, Amotz Zahavi, William Hamilton.

**Evolution of beauty:** Charles Darwin, Stephen Jay Gould, Daniel Dennett.

[7]Previous studies of sexual selection have been criticized for drawing on the male-centered view in the Victorian era, when Darwin advocated his theory of evolution. More recently, some studies have taken a more proactive view of the role of the female. See [22] for details.

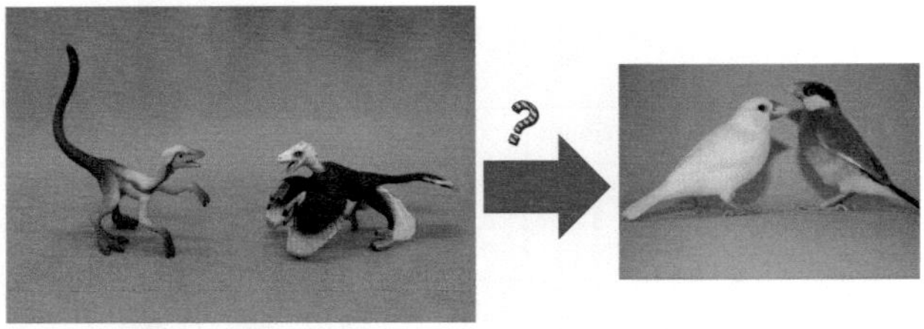

**Figure 2.25:** Feathered dinosaurs; Velociraptor (left), Microraptor (right). They apparently had feathers but could not fly.

Adaptationism is the assertion that the superior fitness will always evolve. Even flamboyant feathers and exaggerated horns, which seem detrimental at first glance, indicate hidden optimality, such as an improved immune system. It is believed to be oriented toward some (as yet unknown) improvement in fitness, even if it is unknown at this time.

Evolution of beauty, on the other hand, is an idea that leads to art for art's sake [103]. It is not always the superior or optimal fit that evolves. In sexual selection, it is rather the female's preference that triggers the runaway effect. The preference in such cases is not always for optimality and may not be suitable for the environment. Gould proposed the concept of exaptation. This is essentially the same as Darwin's pre–adaptation. It is a trait in which one function has evolved by selection to take on another function. Or it can refer to a trait that was not formed as an adaptation to something in the first place, but as a byproduct of another adaptation and now has a different function. An example of this is in the evolution from dinosaurs to birds (see Fig. 2.25). It is believed that bird feathers were originally formed not to fly, but to preserve body heat. It has since been modified by natural selection for the function of flying. Thus, adaptations in evolution are diverted to other uses.

In this sense, Gould criticizes the adaptationist way of thinking (the idea that evolution is viewed as optimization). Gould further points out that it is difficult to distinguish between adaptations (arches) and byproducts of adaptations (spandrels), using the example of the spandrel, a triangular structure sandwiched between arches in Gothic architecture (see Fig. 2.26). Spandrels can have beautiful religious paintings on them, though at first glance they are stunning components. In the first place, it is an extra structure that is created to compose the arch structure. This extra area was used by religious painters [38].

The same idea is found in "competence without comprehension" by Daniel Dennett[8] [29]. He states that the competence that evolution gives to organisms

[8]  Daniel Dennett (1942–): American philosopher, author, and cognitive scientist. Specialized in philosophy of mind, philosophy of science, and philosophy of biology. He is also famous for his explanation of the frame problem in artificial intelligence, using the example of a robot.

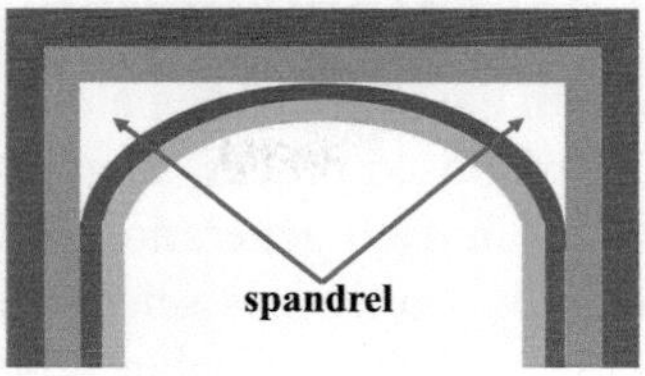


(a) Triangular structures

(b) Gothic architecture (Basilica di San Pietro in Vaticano)

**Figure 2.26:** Byproducts of adaptations in evolution.

(a) Termite mound. The inside of the mound is always maintained at a constant temperature.

(b) Gaudi's Sacrada Familia, so famous as an unfinished work of the architectural genius Antoni Gaudi, is scheduled to be completed in 2026.

**Figure 2.27:** Competence without comprehension.

is not accompanied by comprehension. Natural selection itself is very capable of producing designs of outstanding skill and efficiency without any semblance of understanding. As an example, bacteria and trees have a variety of usefulness (to themselves and to others), but they do not need to understand that usefulness. Consider, for example, the contrast between termite mounds and Antoni Gaudi's Sacrada Familia (see Fig. 2.27). The two are very similar, but there are significant differences in their origins and construction methods. Termite mounds exist for a reason, but none of the termites represent those reasons. This is the competence without comprehension. On the other hand, Gaudi's building also has reasons for its existence, but those reasons are the obvious ones that Gaudi embraced.

## 2.5 Sexual selection for evolutionary computation

Goh [39] introduced sexual selection into GA with the idea that females (males) are in charge of exploration (exploitation). The proposed process is as follows. The population is divided into males and females in each generation. Whereas all females could mate, only those males selected by tournament selection could mate with females. This method is said to allow for a proper balance between exploration and exploitation without special tuning, since the genes of all females remain, but the performance of males is biased.

We will explain comparative experiments with standard GA and sexual selection. The experimental set-ups below are based on reference [152].

First, let us consider the knapsack problem. This problem is typically NP-complete and is known to be computationally difficult to find an optimal solution. The knapsack problem can be formulated as follows (see Fig. 2.28):

> There are $N$ items, and the weight $w_i$ and the value $p_i$ of each item are known. The knapsack has a weight limit $W$, and the weight of items cannot exceed it. Find a method to decide what items to put in the knapsack, such that the sum of the values is maximized.

More formally, the goal is to find $T$, which is a subset of $\{1,2,\cdots,N\}$, that which maximizes $\sum_{i\in T} p_i$, under the following condition:

$$T \subseteq \{1,2,\cdots,N\}, \tag{2.13}$$

$$\sum_{i\in T} w_i \leq W. \tag{2.14}$$

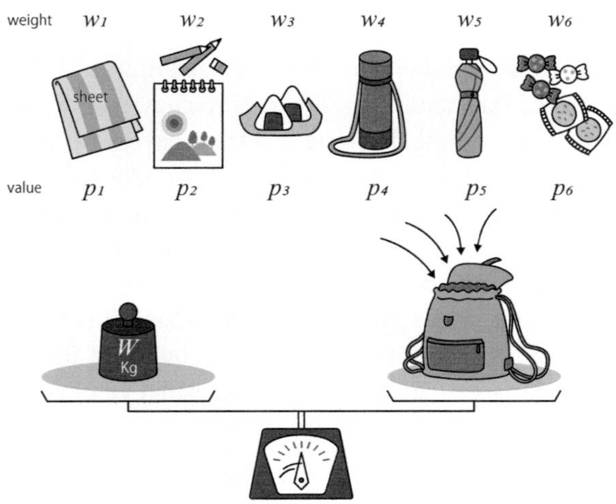


**Figure 2.28:** Knapsack problem.

**Table 2.6:** Knapsack problems.

| **Promblem** | | **Value** |
|---|---|---|
| P07 | weights | 70, 73, 77, 80, 82, 87, 90, 94, 98, 106, 110, 113, |
| | values | 115, 118, 120, 135, 139, 149, 150, 156, 163, 173, |
| | | 184, 192, 201, 210, 214, 221, 229, 240 |
| | capacity | 750 |
| | optimal value | 1458 |
| | optimal packing | 1, 0, 1, 0, 1, 0, 1, 1, 1, 0, 0, 0, 0, 1, 1 |
| P08 | weights | 382745, 799601, 909247, 729069, 467902, 44328, 34610,698150, 823460, 903959, 853665, 551830, 610856, 670702,488960, 951111, 323046, 446298, 931161, 31385, 496951,264724, 224916, 169684 |
| | values | 825594, 1677009, 1676628, 1523970, 943972, 97426, 69666,1296457, 1679693, 1902996, 1844992, 1049289, 1252836,1319836, 953277, 2067538, 675367, 853655, 1826027,65731, 901489, 577243, 466257, 369261 |
| | capacity | 6404180 |
| | optimal value | 13549094 |
| | optimal packing | 1, 1, 0, 1, 1, 1, 0, 0, 0, 1, 1, 0, 1, 0, 0, 1, 0, 0, 0, 0, 0, 1, 1, 1 |

For this problem, the genotype is a set of binary values for choosing (1) or not choosing (0) an item in the knapsack. The population is then divided into males and females. In this method, For each individual, the role of male or female rotates from generation to generation. First, females are selected according to their fitness. When a female (F) is selected, her mating male is selected according to the following criteria:

1. Select the male with the largest Hamming distance from F.
2. If more than one male is selected above, select the male with the best fitness.
3. If multiple males are selected above, select the male with the highest number of 1s.
4. If there is still more than one, select one male at random.

For standard GAs, tournament selection is used. In addition, an acceptance threshold $t$ is introduced and a random number $r$ ($0 \leq r \leq 1$) is generated. When $r \leq t$, the individual with the higher fitness is selected; otherwise, the individual with the lower fitness is selected. In the following experiments, we compare simulations with five levels of acceptance threshold, i.e., 0.75, 0.85, 0.90, 0.95, and 1.0.

For the GAs with sexual selection, the current population is first randomly divided into males and females. Then, for each female, males were selected to mate and produce offspring. The following four selection methods for males were used and compared:

1. Select the male with the highest fitness in a tournament system (without introducing female preferences).
2. Introduce a genotype expressed as a sequence of 0s or 1s in the male as the female's preference. For the selection, the number of 1's in the genotype of the male should be odd.
3. Select the male that chooses the item of greatest value.
4. Select the male that chooses the item of the minimum value.

In the following experiments, we use P07 and P08 problems.[9] Note that the optimal solution in both problem settings includes both the minimum and maximum value objects, and the number of objects chosen is even (see Table 2.6). Therefore, for preferences 2–4, a certain kind of heuristics is included for solving the task.

If the female has a preference, a gene representing that preference is added. Depending on the expression of that preference gene, the female selects a male based on Kirkpatrick's model. When a female expresses a preference (P1), a male (T1) expressing a preference that matches the female's is selected using eq. (2.7). Similarly, when the female does not express a preference (P0), the male (T0) that does not express a preference that matches the female's is chosen by eq. (2.8). In the following experiment, we set $a_0 = 2.5, a_1 = 2.0$. We performed 30 runs of GA with each selection method using the same initial population in two settings, one with 30 individuals evolving for 300 generations and the other with 50 individuals evolving for 500 generations. We compared the performance using the best solution obtained and its average.

Figure 2.29 shows the experimental results on knapsack problems P07 and P08. The figure shows the difference between the optimal value and the evolved solution (the smaller the better) for the different selection methods. The optimal solution evolved in the case of sexual selection with minimal preferences (see Fig. 2.29(a)). On the other hand, Fig. 2.29(b) shows that the optimal solution evolved for tournament selection and sexual selection with odd number of preferences, but the difference for each selection method is small. In P08, sexual selection with preferences is better than that without preferences (Figs. 2.29(c) and (d)). However, the overall trend seems to be that tournament selection gives results closer to the optimal solution than sexual selection. Designing more effective preferences is an interesting topic for future research.

As another example, let us consider $N$–queen problem. The $N$–queen problem is also an NP-complete problem and is defined as follows. On an $N \times N$

[9]Data for the 01 knapsack problem available from `https://people.sc.fsu.edu/~jburkardt/datasets/knapsack_01/knapsack_01.html`.

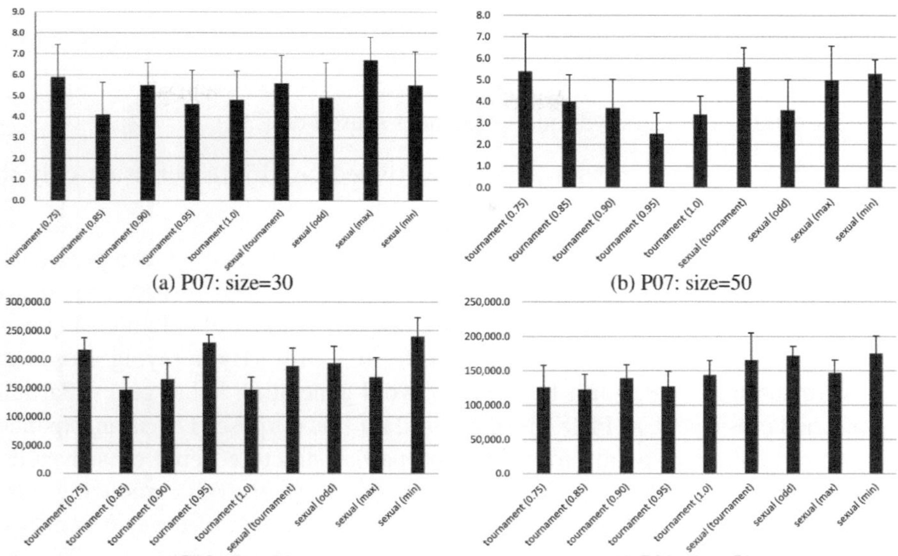


(a) P07: size=30 (b) P07: size=50

(c) P08: size=30 (d) P08: size=50

**Figure 2.29:** Experimental results on knapsack problems.

**Figure 2.30:** 8–queen problem.

square chess board, place $N$ chess queens on each square. Chess queens can move vertically, horizontally, and diagonally as far as they wish. The objective of this problem is to find a game in which no two queens are in the same row, column, or diagonal line (see Fig. 2.30). As $N$ gets larger, it becomes more difficult to generate solutions. Originally it was an 8-queen problem ($N = 8$) and there are 92 unique solutions It is said that the famous mathematician Carolus Fridericus Gauss got it wrong.

We compare the following three types of GAs:

1. Standardl GA

2. Sexual selection without preferences (based on [39])

3. Sexual selection with preferences (based on [152]), but using only the maximum criterion.

Hamming distance (i.e., number of different genes) is used as a preference. These GAs are described as GA, GA+SS, and GA+SS+P, respectively, below. For GA genes, the number of queens is simply the number of columns of queens. Thus, for $N$–queen, it is a permutation from 1 to $N$. By applying crossovers and mutations,[10] care must be taken to avoid generating lethal genes. Partially matched crossovers, or PMXs is often used for this purpose. This is an expanded version of the two-point crossover. The partial structure being crossovered is maintained, but other structures are changed so as to avoid producing lethal genes. It can be explained as follows: Suppose the arrays of the two genotypes are $A[1], A[2], \cdots A[n]$ and $B[1], B[2], \cdots B[n]$ where $n$ is the gene length. Then the PMX crossover is performed by following the procedure:

**Step 1** Two crossover points $i$ and $j$ ($1 \leq i \leq j \leq n$) are randomly selected. In **Step 4**, the section $A[i], A[i+1], \cdots A[j]$ is replaced by the section $B[i], B[i+1], \cdots B[j]$.

**Step 2** When there are common elements in $\{A[1], A[2], \cdots A[i-1], A[j+1], \cdots A[n]\}$ and $\{B[i], B[i+1], \cdots B[j]\}$, let them be $A[k]$ and $B[m]$. At this point, the value of $A[k]$ will be changed to that of $A[m]$.

**Step 3** When there are common elements in $\{B[1], B[2], \cdots B[i-1], B[j+1], \cdots B[n]\}$ and $\{A[i], A[i+1], \cdots A[j]\}$, let them be $B[k]$ and $A[m]$. At this point, the value of $B[k]$ will be changed to that of $B[m]$.

**Step 4** The segment $A[i], A[i+1], \cdots A[j]$ will be replaced by the segment $B[i], B[i+1], \cdots B[j]$.

For example, let's consider the following two genotypes:

**P1** 3 4 9 10 6 8 5 1 7 2

**P2** 2 10 6 5 1 7 4 9 8 3

The PMX execution example would be as described below:

**Step 1** Let us suppose that $i = 5$, $j = 7$ have been selected as the crossover points. In other words, the $A[5] = 6$, $A[6] = 8$ and $A[7] = 5$ of **P1** would be replaced by the $B[5] = 1$, $B[6] = 7$ and $B[7] = 4$ of **P2**.

[10]Swam mutation is often used as a mutation. This exchanges the positions of two randomly selected numbers.

**Step 2** Since we now have $A[8] = B[5] = 1$, the value of $A[8]$ will be replaced by that of $A[5]$, or 6. Similarly, $A[9]$ and $A[2]$ will be replaced by $A[6]$ and $A[7]$, respectively. Replacing them all, **P1** will be as follows:

**P1′** 3 5 9 10 6 8 5 6 8 2

**Step 3** Since we now have $B[3] = A[5] = 6$, the value of $B[3]$ will be replaced by that of $B[5]$, or 1. Similarly, $B[9]$ and $B[4]$ will be replaced by $B[6]$ and $B[7]$, respectively. Replacing them all, **P2** will be as follows:

**P2′** 2 10 1 4 1 7 4 9 7 3

**Step 4** The segment $A[5] = 6, A[6] = 8, A[7] = 5$ will be exchanged with $B[5] = 1, B[6] = 7, B[7] = 4$. The GTYPE of the offspring generated as a result will be:

**C1** 3 5 9 10 1 7 4 6 8 2

**C2** 2 10 1 4 6 8 5 9 7 3

Figure 2.31 shows the experimental results for $N = 10$ with a population of 1,000. For each generation, the average fitness (orange), best fitness (red), count of feasible solutions (blue), and accumulated number of unique solutions discovered ( green) are plotted. The experiment was repeated ten times Since the results were almost the same, a typical example is shown in the figure. We have observed that the standard GA converges the fastest, but with a small number of unique solutions. GA+SS gives the most unique solutions (about 200), although the trend of increasing fitness is inferior. This indicates a concentration on exploration at the expense of exploitation. GA+SS+P evolves about 30 unique solutions, but not as many as GA+SS. Although they eventually converge, the curve of the search oscillates up and down, indicating that the deviation is the largest. In this sense, GA+SS+P is the most balanced of the three methods in terms of exploration and exploitation.

Figure 2.32 compares the average fitness and the number of unique solutions found for different $N$. The figure shows that standard GA is superior in terms of fitness, but that GA with SS is preferable for the search for a variety of solutions.

The above experimental results show that sexual selection guarantees the robustness and solution diversity. In other words, it can balance exploration and exploitation without parameter tuning. However, according to the overall results, the best selection method depends on the problem settings, which is an important future research topic for the practical application of sexual selection.

## 2.6 Novelty search

Novelty search is one of the selection methods proposed by Lehman and Stanley [78, 130]. In conventional evolutionary computation, individuals are evaluated based on their "fitness," which indicates how well they perform on a given

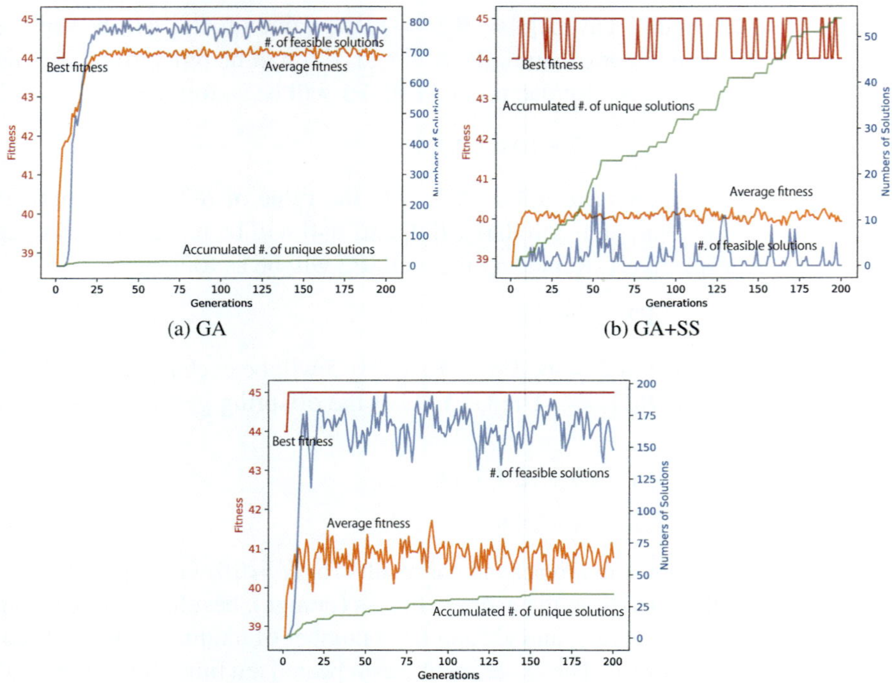


(a) GA

(b) GA+SS

(c) GA+SS+P

**Figure 2.31:** *N*–queen results.

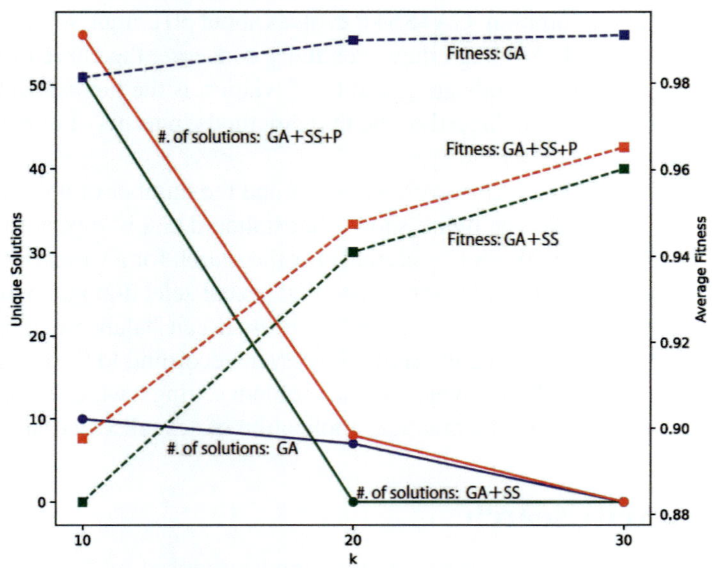


**Figure 2.32:** Knapsack problem.

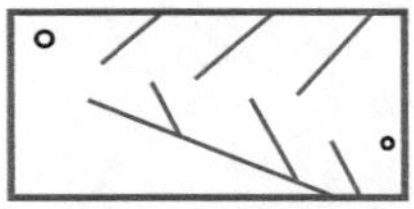

(a) Medium map

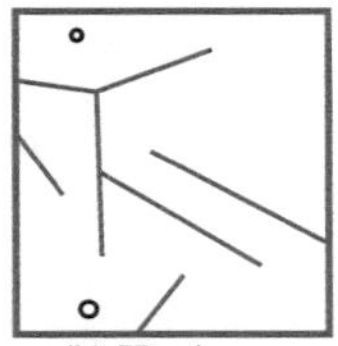

(b) Hard map

**Figure 2.33:** Maze navigation maps.

problem. On the other hand, novelty search introduces the concept of "novelty," which indicates how new an individual is compared to previously discovered individuals, and evaluates individuals based on their novelty. To calculate novelty, it is common to define a "behavior characterization vector (BC vector)" that represents the characteristics of the behavior of an individual, and to calculate the distance between individuals. Specifically, novelty is calculated as follows:

$$nevelty(a) = \frac{1}{k}\sum_{i=1}^{k} dist(a, b_i), \tag{2.15}$$

were $a$ and $b$ are the behavior vectors of individuals in the population, and $dist$ is the proximity (distance) between the two individuals. Here, we are looking for the average distance of the $k$ individuals that are closest (similar) to individual $a$. Intuitively, the novelty is higher for the individual that is farthest from the other individuals.

In addition, a data structure called "archive" is prepared to record good solutions found in the past, and the archive is updated when new good solutions are found. Novelty search computes novelty by referring to the archive, not only to the current population. This avoids falling into the trap of local solutions.

Lehman et al. empirically verified that novelty search guarantees population diversity by using maze problems. They compared novelty search and ordinary evolutionary search in the maze shown in Fig. 2.33. Here, the large circle is start and the small circle is goal. The results are shown in Fig. 2.34. This figure gives a typical run. Each black dot is the end position of the search in the middle of the run, indicating that the novelty search is a broader search than the fitness-based evolutionary computation (the black area extends over the whole area).

Gomes et al. [41] proposed two methods using novelty search to avoid premature convergence in competitive co-evolutionary algorithms.

The first method replaces the conventional evaluation criterion of fitness for all individuals of a population with novelty for some or all individuals of the population.

The second method is called PMCNS (Progressive Minimal Criteria Novelty Search), which is an extension of the novelty search. This method is based on the usual novelty search, but differs in that it defines a "minimal criterion" as the minimum fitness that an individual must satisfy. Individuals whose fitness

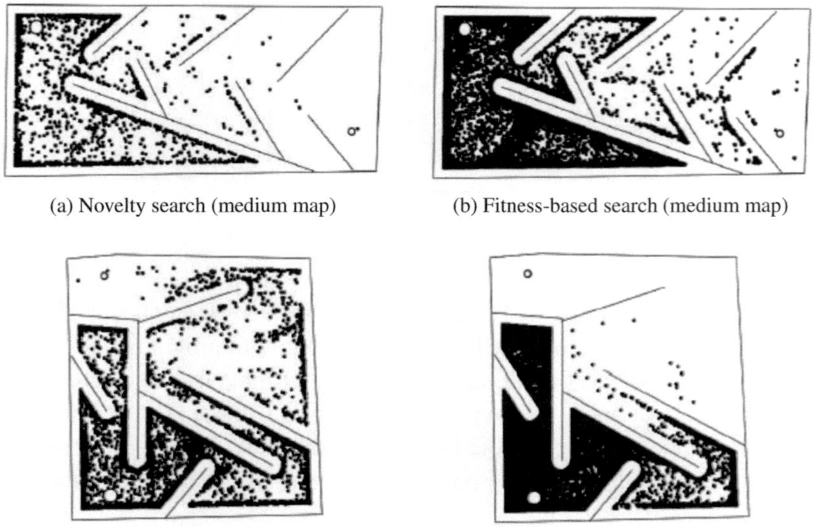

(a) Novelty search (medium map) (b) Fitness-based search (medium map)

(c) Novelty search (hard map) (d) Fitness-based search (hard map)

**Figure 2.34:** Final points visited over typical runs [78].

exceeds the minimal criterion are evaluated as novelty values (non-negative). For individuals whose fitness is less than the minimal criterion, the value is set to be 0. The minimal criterion is defined as the median fitness value of individuals in the population, and the minimal criterion increases as the evolution progresses (in this sense, it is "progressive").

These proposed methods were experimentally compared with conventional methods that evaluate all individuals in the population based on their fitness. The experimental environment was a predator-prey task. The results showed that the PMCNS outperformed the conventional method in the predator evolution, while the conventional method outperformed PMCNS in the prey evolution.

Sexual selection is similar to novelty search in the sense that novelty is discovered. In particular, the run away effect stimulates the discovery of novelty, but afterwards, some restrictions prevent the expression of new traits, and the discovery of novelty tends to stagnate. Hintze et al. [51] experimentally verified this by comparing sexual selection and novelty search. The results showed that sexual selection searches more effectively than novelty search in the presence of gene recombination. In other words, recombination is important for the runaway effect, and genes must be exchanged between the two sexes. When this constraint is satisfied, sexual selection is shown to find novelty more efficiently than novelty search.

Let us compare novelty search and sexual selection for the dynamic knapsack problem. Here, we use the dataset P07 (see Table 2.6). In this experiment, the knapsack capacity changes during the evolution. Figure 2.35 shows the averages of the best fitness (i.e., total values of the items in the knapsack) for each generation over 30 runs. We have chosen experimental parameters of 30 individuals, 500 generations, and a mutation rate of 0.01. As can be seen from the figure, the capacity of the knapsacks changed in generations 50, 150, 250, 350 and 450.

First, we made comparative experiments of tournament selection with different elite rates (i.e., 75%∼100%) (see Fig. 2.35(a) and (b)). The higher the elite selection rate, the greater the drop in fitness when the knapsack capacity is reduced and the longer it takes to recover. This is because, in case of high elite rates, more individuals with high fitness values are left in the next generation, thus narrowing the search space and making it easier to fall into local solutions. As a result, when the environment changes rapidly, diverse individuals will be lacking and the search for novel solutions will be delayed.

As for the sexual selection, the following four different methods were compared with respect to male attractiveness (see Fig. 2.35(c) and (d)):

**Tournament:** All males have the attractive characteristics equal to the total values of the items in the knapsack.

**Odd:** Individuals choosing the odd number of items have the attractiveness.

**Max:** Individuals choosing the item with the highest value have the attractiveness.

**Min:** Individuals choosing the item with the lowest value have the attractiveness.

The results of the experiment showed little difference in performance. However, compared to tournament selection, there was less of a drop in fitness values and relatively quicker recovery when knapsack capacity was reduced. This is because the sexual selection methods can quickly cope with environmental changes while maintaining diversity. In particular, the ability to maintain broad genetic diversity allows for more adaptive search for dynamic environments.

On the other hand, novelty search has the excellent feature of searching a wide search space (Fig. 2.35(e)). The drop in fitness transition was relatively small when the knapsack capacity was reduced. However, its subsequent performance improvement was inferior to that of the other methods. This is because novelty search focuses on "novelty" rather than fitness, which tends to slow the speed of convergence toward the optimal solution. In other words, while novelty search is effective for avoiding local solutions, it may not be suitable for the purpose of improving the fitness values explicitly.

MAP-Elites (Multi-dimensional Archive of Phenotypic Elites) is an evolutionary algorithm inspired by novelty search [89]. Map-Elites selects some solution features of interest in the problem and exhaustively calculates how the

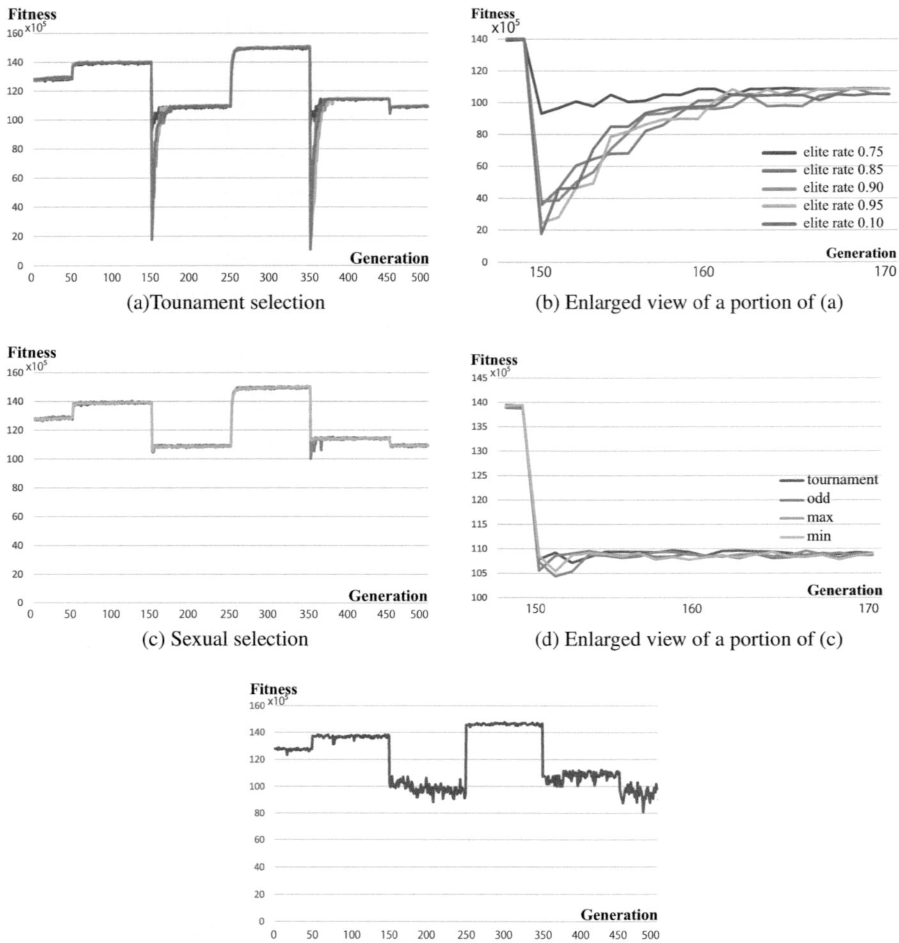


(a)Tounament selection

(b) Enlarged view of a portion of (a)

(c) Sexual selection

(d) Enlarged view of a portion of (c)

(e) Novelty search

**Figure 2.35:** Solving a dynamic knapsack problem.

adaptations (i.e., fitness values) are distributed in that feature space. The feature space is divided into cells of a certain granularity, and evolutionary computation proceeds by selecting the individual with the best fitness value (i.e., elite) within each cell. Since elites are derived for each cell, it is possible to maintain higher diversity than in novelty search. It has been reported that relatively high fitness values can be obtained as a result of evolution.

Note that MAP-Elites seeks only a single best solution ("occupant") for each cell. Thus, if a newly generated solution belongs to a cell that already has an occupant, the solution with a higher fitness value compared to that occupant remains as the next occupant. This property reflects the fact that in nature, selection occurs through local competition with neighboring individuals.

MAP-Elites works in the following steps:

**Step1** Generate random initial solutions (genotypes).

**Step2** Map the generated solution population to the cells in the feature space and leave only the occupants.

**Step3** Apply mutation to a solution randomly selected from the solution population to generate a new solution.

**Step4** Calculate the features and the fitness value of the new solution.

**Step5** If the fitness is higher than that of the occupant of the cell to which the solution belongs, it is replaced.

**Step6** Repeat **Step3** to **Step5**.

Note that not all cells always have a solution, since there may be cases where no solution exists within the range of the feature values corresponding to each cell. In addition, there is a possibility that a solution cannot actually be explored even if a solution in that cell exists.

In other words, MAP-Elites is an algorithm that tries to find the best-performing solution in the solution space while maintaining diversity by searching for solutions in each region of the feature space. As an example of MAP-Elites, CVT-MAP-Elite has been proposed [153], which partitions the feature space using Centroidal Voronoi Tessellation (CVT).

# Chapter 3

# Swarm Intelligence Approach to AI

While the ants worked hard, the grasshopper did nothing but play and sing. Finally, the cold winter arrived. The grass in the field was completely dried up, and not a single thing was left for the grasshopper to eat. (quoted from "The ant and the grasshopper", Aesop's Fables).

Search methods based on biological or physical phenomena are called "metaheuristics." GA, GP, Simulated Annealing (SA), neural networks are reinforcement learning are also metaheuristics in a broad sense, because they are based on evolutionary theory, physical phenomena, brain plasticity, behaviorism, and so on.

This chapter provides an overview of metaheuristics and recent research results. Criticisms of metaheuristic research are also discussed. In addition, we experimentally examine swarm intelligence in ants and the latest metaheuristic research using it. Improvement of metaheuristics using generative AI and generation of novel metaheuristics will also be explained.

## 3.1 An overview of metaheuristics and critics

Metaheuristics are broadly classified as follows (see Fig. 3.1):

- Evolutionary computation
- Swarm intelligence

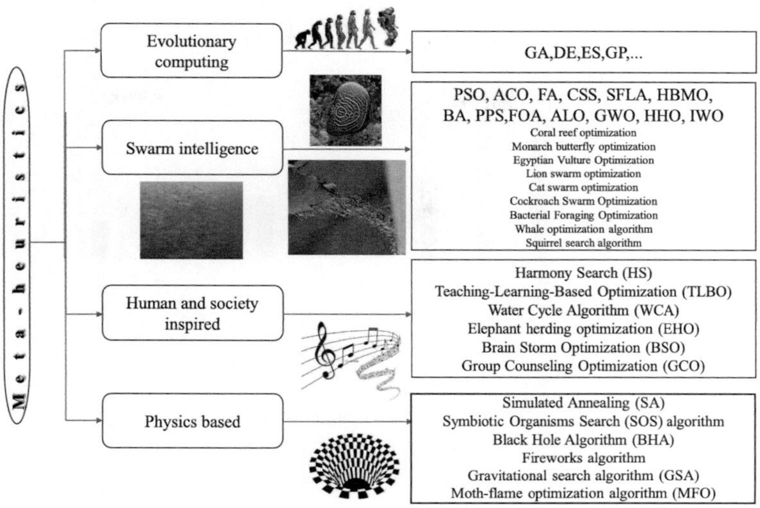


**Figure 3.1:** Overview of meta-heuristics.

- Human and society inspired methods
- Physics based mathods

These metaheuristics methods are frequently applied to various areas. Figure 3.2 shows the frequency of literatures published in top journals 3.2. However, the difference in frequency does not necessarily indicate validity. It seems to depend on clarity, ease of implementation, and applicability of the method.

On the other hand, criticism has been thrown at this seemingly excessive variety of metaheuristics. We will explain this based on one example.

Consider harmony search [36], which is a metaheuristic method based on jazz session (generation process of human improvisation). Musicians are supposed to perform improvisation mainly using any one of the methods as outlined below:

- Use already-known scales (stored in their memory).
- Partially change or modify the already known scales. Play scales next to the one stored in their memory.
- Create new scales. Play random scales within their playable area.

A process whereby musicians combine various scales in their memory for the purpose of composition is regarded as a sort of optimization. While many metaheuristics are based on swarm intelligence of life such as fish, insects, etc., harmony search (referred to as “HS,” hereinafter) significantly differs from them in terms of exploiting ideas from musical processes to search for harmony, according to an aesthetic standard.

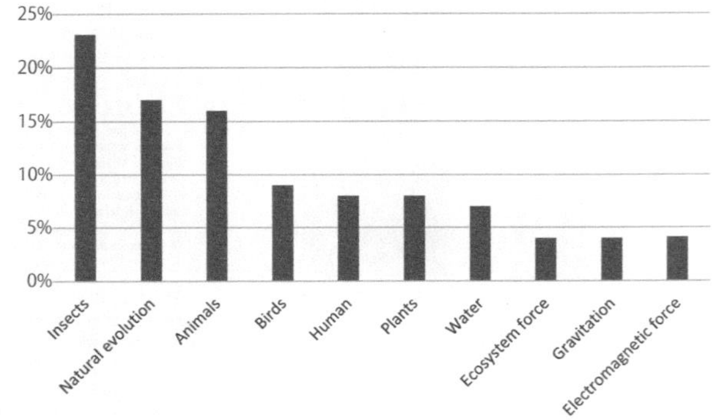


(a) Top ten leading areas.

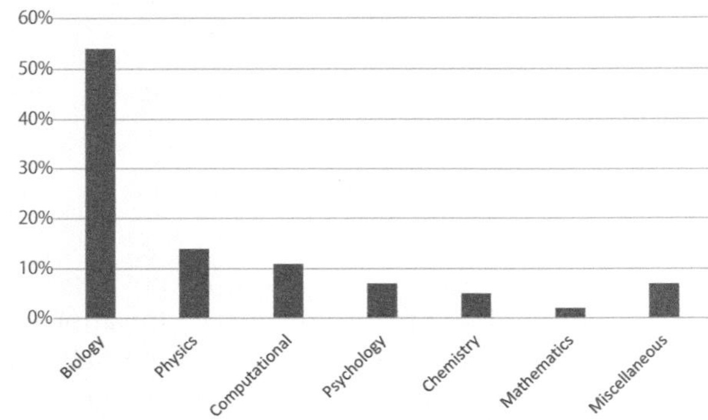


(b) Top six leading disciplines.

**Figure 3.2:** Which areas of the metaheuristic are frequently used?

Harmony search algorithms search for the optimal solution by imitating a musician's processes according to the following three rules:

- Select an arbitrary value from HS memory.
- Select a value next to the arbitrary one from HS memory.
- Select a random value within a selectable range.

With HS, a solution candidate vector is referred to as a harmony, and a set of solution candidates is referred to as a harmony memory (HM). Solution candidates are replaced within HM by a specific order. This process is repeated a certain number of times (or until the conditions for termination are met), and finally, the best harmony is selected among those that survive in HM as a final solution.

**Algorithm 3** shows Harmony Search algorithm. *HMCR* (Harmony Memory Considering Rate) represents the probability of selecting a harmony from HM, while *PAR* (Pitch Adjust Rate) represents the probability of amending a harmony selected from HM. *HMS* is the number of harmonies (sets), which is normally set to between 50 to 100.

A new solution candidate (harmony) is generated from HM based on *HMCR*. *HMCR* is the probability of selecting component elements[1] among the present HM. Thus, new elements are randomly generated by the probability of $1-HMCR$. Subsequently, mutation occurs according to the probability of *PAR*. The *bw* parameter (bandwidth) represents the largest size of the mutation. In case a newly generated solution candidate (harmony) is better than the poorest solution of HM, it is replaced.

This method is similar to a genetic algorithm (GA); however, it differs in that all the members of HM become a parent candidate in HS while only one or two existing range(s) of chromosomes (parent individuals) are used to generate a child chromosome in GA.

Some critical opinions have been expressed against meta-heuristics and further discussions have also been undertaken.

Meta–heuristics frequently uses unusual names, terms associating with nature, and metaphors. For example, in the harmony search, the following terms are used:

- harmony
- pitch note
- sounds better

However, these are just saying the following words listed below in another way:

- solution
- decision variable
- has a better objective function value

Although critics insist that these replaced words cause confusion [126], it is considered that use of metaphorical expressions itself does not influence the ease of understanding. For example, David Hilbert[2] is quoted as saying Geometry does work even if a point, line and face are expressed as a table, chair, and beer mug in discussing mathematical forms. That is to say, it does not matter at all in precise

[1] Component element corresponds to allele in genotype under GA.

[2] 
David Hilbert (1862–1943): German mathematician. At the 2nd International Congress of Mathematicians (ICM) in Paris in 1900, he made a speech on "problems in mathematics," where he stressed the importance of 23 unsolved problems and presented a prospect for future creative research through these problems. Some of them continue to be themes for research on mathematics and computer science.

**Algorithm 3** Harmony search.

**for** $i = 1$ to $HMS$ **do** ▷ HM initialization.
  **for** $j = 1$ to $n$ **do** ▷ $n$: harmony length.
    Randomly initialize $x_j^i$ in HM ▷ $x_j^i$: $j$-th position of $i$-th harmony.
  **end for**
**end for**
**while** the stop criterion is not satisfied **do**▷ Generate a new solution candidate $\vec{x}$.
  **for** $j = 1$ to $n$ **do**
    **if** rand(0,1)$<$ $HMCR$ **then**
      Let $x_j$ in $\vec{x}$ be the $j$-th dimension of a randomly selected HM member
      **if** rand(0,1)$<$ $PAR$ **then**
        Apply pitch adjustment distance $bw$ to mutate $x_j$
        $x_j = x_j \pm \text{rand}(0,1) \times bw$
      **end if**
    **else**
      Let $x_j$ in $\vec{x}$ be a random value
    **end if**
  **end for**
  Evaluate the fitness of $\vec{x}$ by $f(\vec{x})$
  **if** $f(\vec{x})$ is better than the fitness of the worst HM member **then**
    Replace the worst HM member with $\vec{x}$ ▷ HM update
  **else**
    Disregard $\vec{x}$
  **end if**
**end while**
Postprocess results and visualization

discussions when it is axiomatically defined. Nevertheless, we should be careful about insisting on the novelty of meta-heuristics. It is important to recognize the distinct difference with existing methods for further discussions.

Wayland [163] has criticized the Harmony Search as simply a special example of evolution strategy $(\mu+1)ES$ (see **Algorithm 4**), which is one of the evolutionary algorithms developed by Ingo Rechenberg [109].[3] Comparing this algorithm to that of the harmony search (**Algorithm 3**), it is assumed that the essential method of calculation is the same although they slightly differ in parameter setting, among others. On the contrary, discussion on methods of expanding

[3] 
Ingo Rechenberg (1934–2021): German professor in the field of bionics. He was a pioneer of the fields of evolutionary computation method known as evolution strategies from German "Evolutionsstrategie."

---

**Algorithm 4** Evolution Strategy: $(\mu+1)ES$.

Initialize the population with randomly generated solutions
**while** the stop criterion is not satisfied **do**
  Create a new solution using recombination and mutation operators
  **if** the new solution is better than the worst solution in the population **then**
    Replace the worst solution by the new one
  **end if**
**end while**
Return the best solution in the population
Postprocess results and visualization

---

the harmony search algorithm and the parameter setting methods can be built up in a different way to ES.[4]

It does matter that researchers who propose meta-heuristics do not recognize other similar methods [126]. When developing a new method, we never fail to have discussions on the basis of past investigations. There is a well-known Chinese proverb which states "Learn from the past and know the new" by Confucius.[5] Returning to an old method and expanding it to a produce a new one is not a bad thing and is in fact recommended. Especially, in the pursuit of research into rapidly developing AI, long-forgotten technologies are expected to revive with these new technologies (big data, supercomputers, among others), and many examples are well known. Now is the time to recognize the importance of engaging in productive discussions with a particular emphasis on old methods. The expectation in right-minded research communities is to allow various proposed methods to compete against each other, which should result in the continuation of more robust approaches for future generations after a process of successive elimination.[6]

## 3.2 Ants know Buffon's needle?

Consider dropping an infinitesimally thin needle of length $L$ onto a plane with lines drawn parallel to each other at equal intervals $L$ (Fig. 3.3). Suppose that

---

[4]The counterargument has been made that the above two methods are different. Refer to `https://en.wikipedia.org/wiki/Harmony_search` for further discussion.

[5]  Confucius (BC552–BC479): A philosopher in the Spring and Autumn period in China. He is the founder of Confucianism. The "Analects of Confucius" (The Governing Volume) is a collection of sayings and anecdotes attributed to him and his disciples. For over two thousand years, the ideology of Confucius was respected and inherited as China's legitimate ideology and the idea of "learning from the past to develop new ideas" is attributed to the Analects (Wei Chang book). Confucius has supposedly said that a requisite for someone to be his disciple is to study ideas and academic disciplines from his predecessors.

[6]Those who engage in AI research based on evolution naturally expect evolutional phenomena (e.g., speciation, diversity, natural selection, pre-adaptation, punctuated equilibrium and so on) in research results.

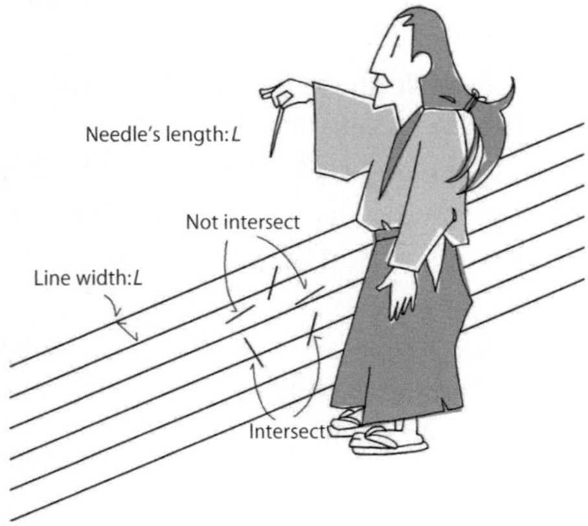


**Figure 3.3:** Buffon's needle.

the needle falls in a random direction and in a random place. The needle may or may not intersect the line. In this case, the probability that the line and needle intersect is $\frac{2}{\pi}$. This problem is known as Buffon's needle. More precisely, if a needle of length $B$ is dropped onto a plane with parallel lines drawn a distance of $I$ apart, the probability $p$ of intersecting the parallel lines is as follows:

$$p = \frac{2B}{I\pi}. \tag{3.1}$$

From this, the area $A$ of a region is inversely proportional to the number of times $N$ that lines of total length $S$ and $L$ dropped randomly on it intersect. In other words, we obtain the following equation:

$$A \approx \frac{2SL}{\pi N}. \tag{3.2}$$

Certain ant species (*Leptothorax albipennis*) are known to move nests. At that time, the scout ants search for several places and decide whether to adopt a certain potential nest site as a new nest. In other words, the decision of the scouts determines whether to move to a nest above a certain standard size or to reject a potential nest that is too small. Mallon et al. [83] have shown that this decision by scout ants is based on Buffon's needle. Specifically, one scouting ant searches once for a burrow while dropping pheromones that can be detected over time, and later searches once more while counting the number of interactions with the pheromones dropped at that time. Based on the information obtained, they estimate the size of their burrows, and when they find a large burrow, they migrate with their mates.

Mallon et al. constructed various environments and observed the search for scout ants. Figure 3.4 shows the actual environments they constructed. Here, (b) and (c) are half the area of a standard nest (a) with the same circumference and

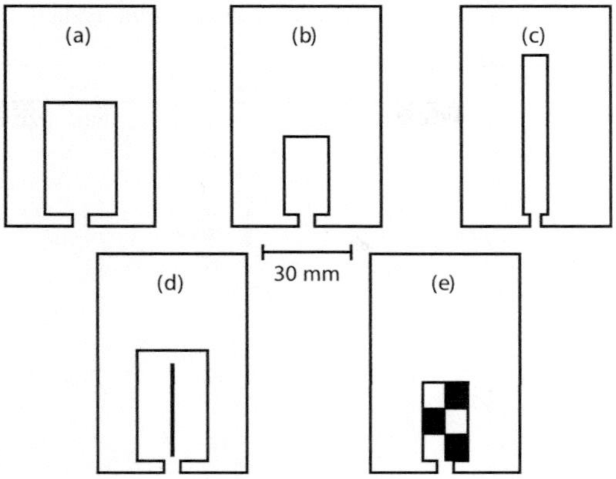


**Figure 3.4:** Ant nest designs [83].

half the area, respectively. In addition, (d) has a wall and (e) has a hole. The results of the experiment showed that 10 out of 13 times the ants chose (a) from (c). This suggests that the ants are making a selection based on Buffon's needle. Table 3.1 shows more detailed experimental data.

Now let us simulate an ant search for this task. This is performed as follows [31, 90]:

1. Let an ant move around randomly over a certain area. Let $L$ be its path length.

2. Next, let the ant move around randomly again with a path length $S$ and record the number of intersections $N$ with the previous path.

3. $K = \frac{2SL}{\pi N}$ is recorded.

4. Repeat the above several times and derive the average of $K$ values to approximate the area.

Approximate values were calculated for various areas and compared to the actual values. In addition, we discussed the validity of the results by comparing the observed selection of real ant nests.

First, we randomly searched within the rectangular area shown in Fig. 3.4 and obtained an approximation of the area based on the Buffon's needle. Then, we checked the consistency against the actual behavior taken by the ants. The area of the nest was determined as follows:

1. Search twice for the path length 3000 and derive the number $N$ of intersections of the two paths.

2. The tentative area of the nest is $\frac{2 \times 3000^2}{\pi N}$.

3. Repeat the above process 300 times and take the average as the area.

**Table 3.1:** Comparison of different nests.

| Experiment | | Nest choice | |
|---|---|---|---|
| A | twice of (a) | (a) | half size of (a) |
| chosen | 7 | 8 | 1 |
| B | (a) | $\frac{5}{8}$ of (a) | |
| chosen | 15 | 0 | |
| C | (a) | (c) | |
| chosen | 10 | 3 | |
| D | (a) | (d) | |
| chosen | 6 | 9 | |
| E | (a) | (e) | |
| chosen | 12 | 8 | |

**Table 3.2:** Results of simulation-based estimation of areas.

| Nest | Real area | Estimated area | Relative error (%) |
|---|---|---|---|
| (a) | 1500 | 1531.7 | 2.12 |
| (b) | 750 | 765.56 | 2.08 |
| (c) | 750 | 785.96 | 4.79 |
| (d) | 1500 | 1611.7 | 7.45 |
| (e) | 750 | 1553.2 | 107 |

Repeatedly, the average value of the area obtained is the final estimate. In (e), however, when the path changed direction on the carpet (black area) during the first search, we constrained the path to not intersect with the second path until the next change of direction. Examples of ant behavior and area estimates for nests A and D are shown in Fig. 3.5. The relative errors between the estimated and actual values of areas are shown in Table 3.2. The table shows that the estimates are nearly correct for (a)–(d). For nest (e), an estimate of about twice the actual area was obtained. This can be explained by the fact that the estimated area is inversely proportional to the number of intersections and that half of the nests were carpeted (black area).

Let us examine the accuracy of nest selection. We consider a situation in which there is one nest hole (a) and one nest hole of another type, and assume that the ant selects the one with the larger estimated area as its emigration site. If the ant can select the larger nest, it is considered to have estimated the size of the nest with sufficient accuracy. The data obtained above are considered to be pairs of data for (a) through (e). For each pair of data, we compared the approximate area of (a) with the approximate area of the other next, and examined the rela-

**Table 3.3:** Selection frequencies of different nests.

| Nest pairs | Selected (a) | Not selected (a) |
|---|---|---|
| (a) and (b) | 986 | 14 |
| (a) and (c) | 935 | 65 |
| (a) and (d) | 481 | 519 |
| (a) and (e) | 513 | 487 |

tionship between the two to see if (a) should be selected. The results are shown in Table 3.3.

The table shows that (b) has half the area of (a), but when compared to (a), (a) was selected with a probability of 98.6%. (c) is also half the area of (a), and the ant can determine that (a) is larger than (c) with a probability of over 90%. There was no significant difference between (a) and (d), and (a) and (e) in terms of selection frequencies. These results are consistent with the actual experimental results of ants shown in Table 3.1. The comparison with (c) and (d) confirms that the information on the circumference and the distance walked before the collision with the wall, respectively, do not affect the area estimate.

Even though the area of (e) is half that of (a), it is selected as much as (a), suggesting that the ants detect the intersections and estimate the area correctly. Furthermore, they are using the fact that the number of their intersections is inversely proportional to their area to make their estimates. Overall, the area estimation by ants using the Buffon's needle principle is sufficiently useful for the purpose of selecting larger nests.

Next, let us experiment with other types of areas. Figure 3.6 shows examples of ant behavior and area estimation for a circle and a star. The radius of the circle is 30 mm and the side length of the star is 14.536 mm.

The estimated area of the circle was 2788.5 ($mm^2$), whereas the actual area is $30^2 \times \pi = 2827.4$ ($mm^2$). This relative error is about 0.01, which means a very good approximation. The estimated area of the star shape, which is non-convex, was 509.1 ($mm^2$). The actual area is 446.7 ($mm^2$).

From these results, it was found that an approximation of the area can be obtained for nonconvex shapes and circles, albeit with somewhat less accuracy.

Future works include experimental testing of other theories, such as,

- Selection based on perimeter,
- Selection based on the longest time to collide with the wall.

Although these are ruled out by real measured results, artificial life simulation experiments would be beneficial. In addition, testing how the number of crossings is recognized by pheromones and the relationship between search time and path length (i.e., $L, S$) would be an interesting research topic.

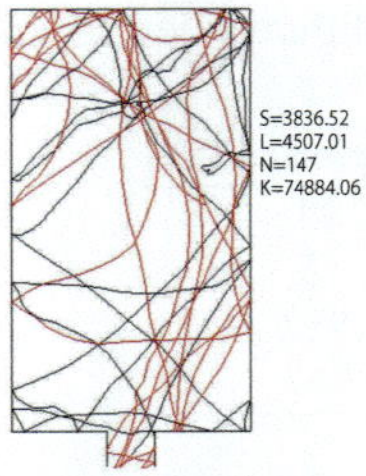


Ant movements for nest (a)

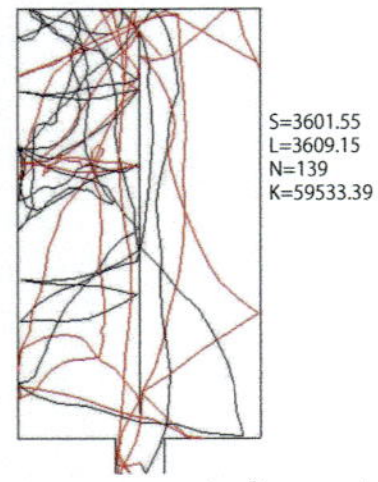


Ant movements for nest (d)

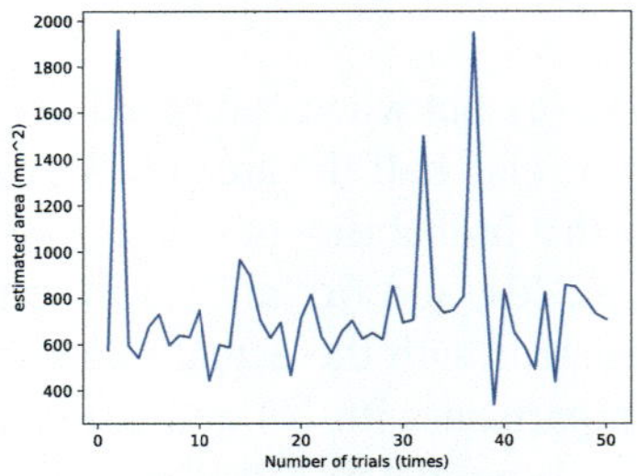


Area estimates for nest (a)

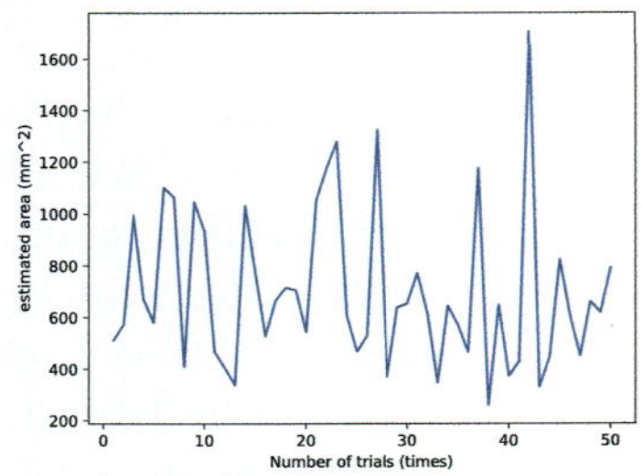


Area estimates for nest (d)

**Figure 3.5:** Ant behaviors and estimation (1).

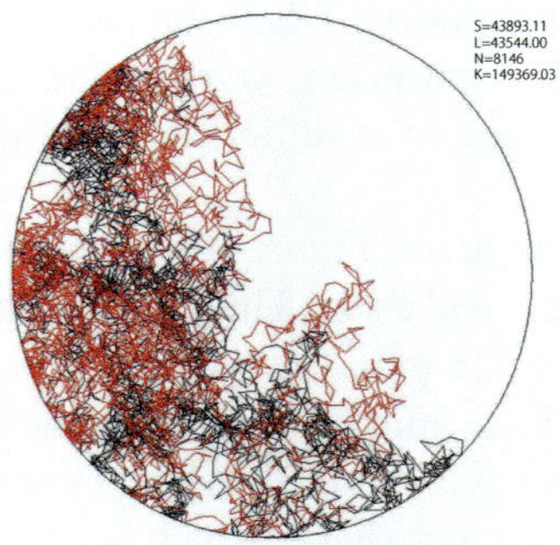


Ant movements for circle

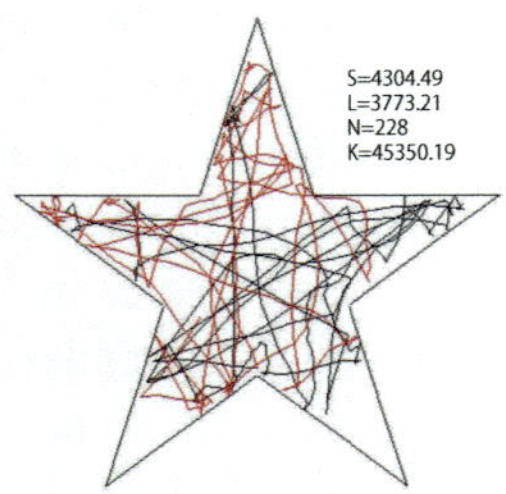


Ant movements for star

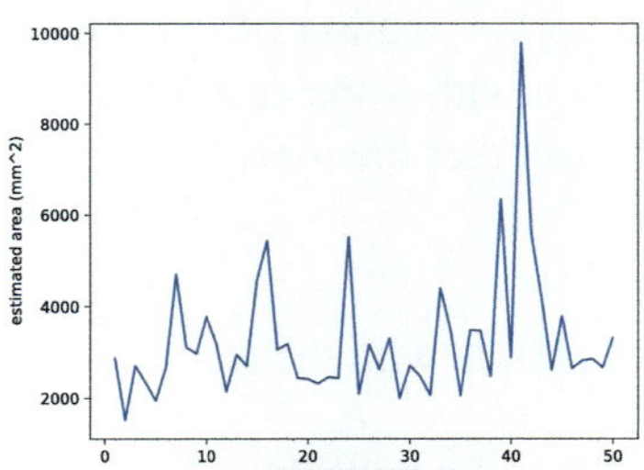


Area estimates for circle

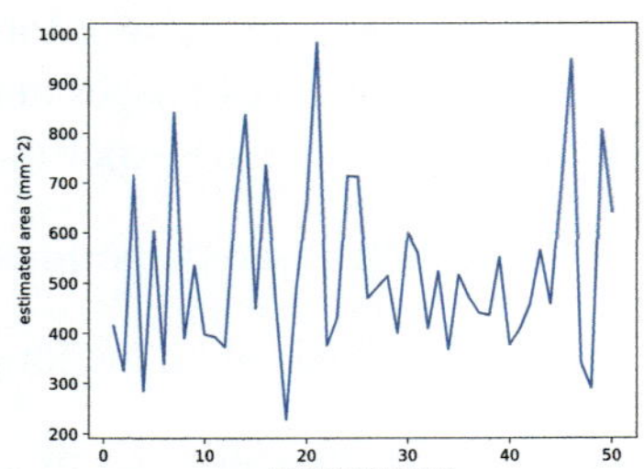


Area estimates for star

**Figure 3.6:** Ant behaviors and estimation (2).

## 3.3 Collective behaviors of ants

Ants march in a long line. There is food at one end and a nest at the other. This is a familiar scene in gardens and on roads, but the sophisticated distributed control by these small insects was recognized by humans only a few decades ago.

Ants established their life in groups, or colonies, more than a hundred million years before humans appeared on Earth. They formed a society that handles complex tasks such as food collection, nest building, and division of labor through primitive methods of communication. As a result, ants have a high level of fitness among species, and can adapt to harsh environments. New ideas including routing, multiple agents, and distributed control in robotics have developed based on simple models of ant behavior. Applications of the ant behavior model have been used in many papers, and are becoming an attractive field of research.

Marching is a cooperative ant behavior that can be explained by the pheromone trail model. Cooperative behavior is frequently seen in ant colonies, and has attracted the interest of entomologists and behavioral scientists. Pheromones are volatile chemicals synthesized within the insect, and are used to communicate with other insects of the same species. Examples are sex pheromones that attract the opposite sex, alarm pheromones that alert group members, and trail pheromones that are used in ant marches.

However, recent research indicates that pheromones are effective within a distance of only about 1 m from the female. Therefore, it is still not known if males are attracted only because of the pheromones.

Many species of ants leave a trail of pheromones when carrying food to the nest. Ants follow the trails left by other ants when searching for food. Pheromones are volatile matter that is secreted while returning from the food source to the nest. Experiments shown in Fig. 3.7 by Deneubourg and Goss using Argentine ants linked this behavior to the search for the shortest path [37]. They connected bridge-shaped paths (two connected paths) between the nest and the food source, and counted the number of ants that used each path. This seems like a simple problem, but because ants are almost blind and they cannot use complex methods to communicate, they have difficulty recognizing junctions or food positions. Furthermore, all the ants must take the shorter path to increase the efficiency of the group. Ants handle this task by using pheromones to guide the other ants.

Figure 3.8 shows the ratio of ants that used the shorter path [37]. Almost every ant used the shorter path as time passed. Many of the ants return to the shorter path, secreting additional pheromones; therefore, the ants that follow the pheromone also take the shorter path. This model can be applied to the search for the shortest path, and is used to solve the traveling salesman problem (TSP; see Section 3.6) and the routing problem of networks. There are many unknown factors about the pheromones of actual ants; however, the volatility of pheromones can be utilized to build a model that maintains the shortest path while adapting to

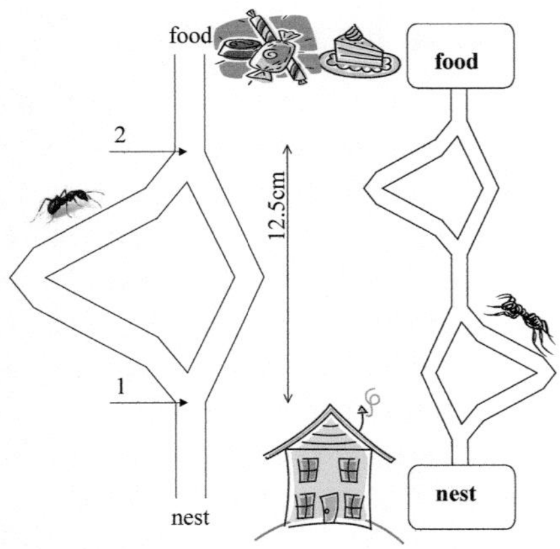


**Figure 3.7:** Bridge-shaped paths.

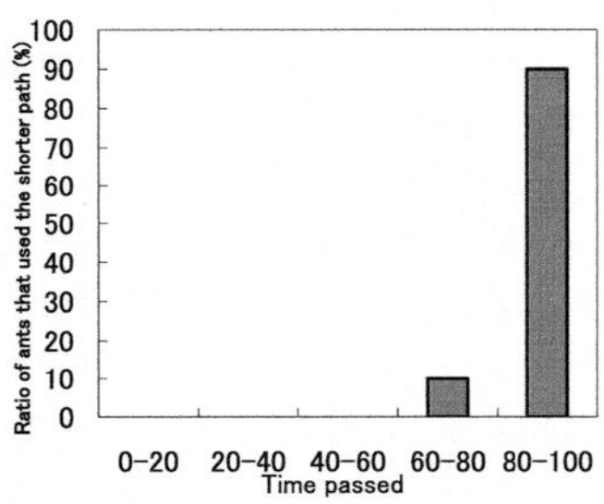


**Figure 3.8:** The ratio of ants that used the shorter path.

rapidly changing traffic. The path with a greater accumulation of pheromones is chosen at junctions, but random factors are inserted to avoid inflexible solutions in a dynamic environment.

An easy model can describe the actions of ants as follows:

- In case of nothing, random search is done.
- If the food is found, takes it back to the hive. Homing ant knows the position of the hive, and returns almost straight back.
- Ants that take the food back to hive drop their pheromone. Pheromones are volatile.
- Ants not having the food have the habit of being attracted to the pheromone.

Figure 3.9 shows the simulation of the pheromone trails. Here, the hives are placed in the center, and there are three (lower right, upper left, lower left) food sources. (a) is the first random search phase. In (b), the closer lower right and lower left food are found, and the pheromone trail is formed. Upper left is in the middle of formation. In (c), pheromone trails are formed for all three sources, and makes the transport more efficient. The lower right source is almost exhaustively picked. In (d), the lower right food source finishes, and the pheromone trail is already dissipated. As a result, a vigorous transportation for the two sources on the left is being done. After this, all the sources finish, and the ants returns to random search again.

Let us try a nasty experiment on ants (Fig. 3.10(a)). It is a cruel game that we used to play as children, but we can perform it without worrying if it is a simulation. First, we let the ants create a pheromone trail (Fig. 3.10(b)). After that, place an obstacle on the trail (Figure 3.10(c)). The pheromone trail is then blocked so that one path (the upper one in the figure) is longer than the other (the lower one). The ants will then become lost (Fig. 3.10(d)), but after a while they will start carrying their food along the shorter path (Fig. 3.10(e)).

This phenomenon is explained by the volatile nature of the pheromone as follows. Probabilistically, ants initially pass through the two paths in the same way. Consider the case where there are two paths to a feeding site, one longer and one shorter. Since pheromone is volatile, the amount of pheromone accumulated in the shorter path is higher. The amount of pheromone that accumulates on the shorter path is greater, and subsequent ants are more likely to follow the shorter path. As a result, over time, almost all ants will take the shorter route. Thus, through pheromone communication, ants are able to stochastically select the best path for the group.

Recent studies [32] have shown that ants utilize pheromones with different volatility levels when foraging for food and when returning home. More specifically, they utilize two types of pheromones:

**Exploration pheromones** Deposited by ants when exploring their environment.

**Foraging pheromones** Deposited by ants to recruit their congeners after the discovery of a food source.

Let us examine how the use of multiple types of pheromones affects the search. In the following simulation, ants are assumed to have three states:

**Exploration state** Ants move randomly to search for a food source. They deposit exploration pheromones along the way.

**Pheromone state** Ants move using information from exploration pheromones to search for a food source. They deposit exploration pheromones on the way.

**Homing state** After ants have found food, they use the foraging pheromones to return home. They deposit foraging pheromones along the way.

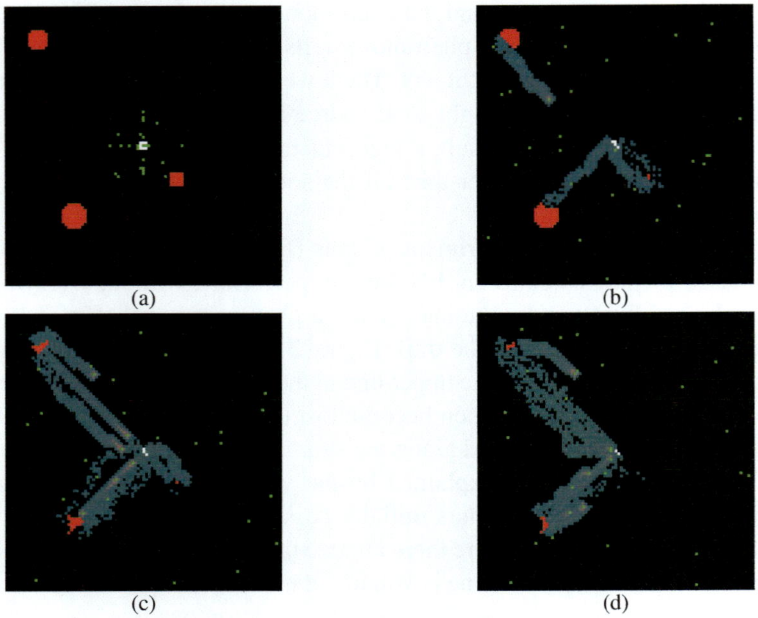


**Figure 3.9:** Pheromone trails of ants.

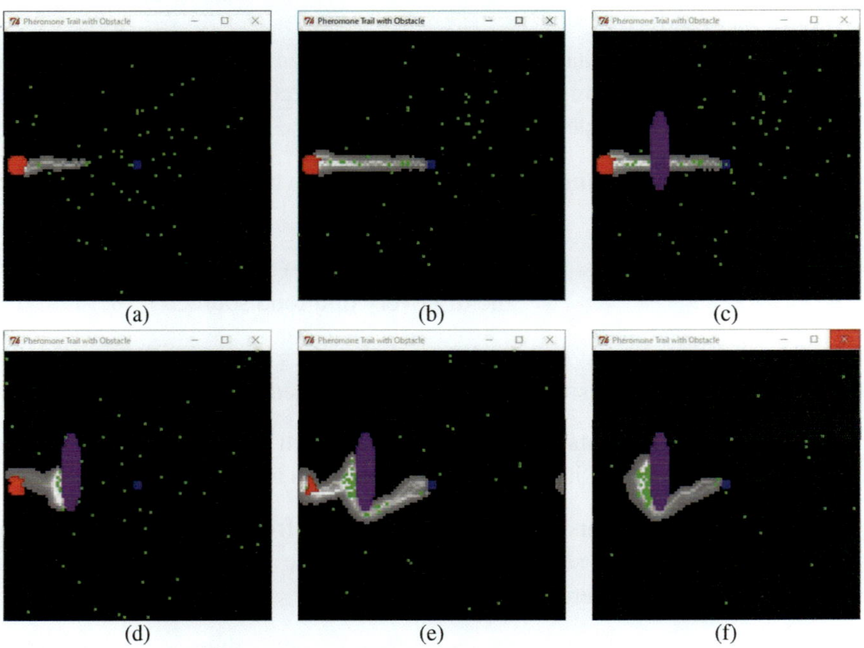


**Figure 3.10:** An obstacle placed in a trail.

When ants move, they determine the direction of movement according to the concentration of pheromones. The higher the concentration, the higher the probability of movement in that direction. In addition, the concentration of pheromones decays exponentially as follows:

$$F = Ae^{-\lambda t}, \tag{3.3}$$

where $\lambda$ is the attenuation rate. According to [32], exploratory pheromones are not as volatile as foraging pheromones and evaporate more slowly. For this reason, a large $\lambda$ value for exploration pheromones and a small $\lambda$ values for foraging pheromones are used in the following experiments. It is expected that more efficient search can be conducted by using different pheromones with different volatility. In other words, the early evaporation of foraging pheromone is aimed at improving overall efficiency. The foraging pheromone will quickly cut off the pathway when foraging is finished and other sources are searched for.

Let us investigate how the presence or absence of exploration pheromones affects search efficiency. Here, we consider the following three strategies for how ants search for a food source when they do not have food:

**Condition A** Ants search for a food source while moving randomly around their surroundings. In this case, they do not deposit the exploration pheromones.

**Condition B** Ants search for a food source while moving randomly around their surroundings. In this case, they periodically deposit the exploration pheromones as they move.

**Condition C** Ants search for a food source while avoiding the exploration pheromones. In this case, they move while periodically depositing the exploration pheromones.

For each condition, search efficiency was compared for the following cases:

- When there is an obstacle between the food source and the nest.
- When the distance between the food source and the nest is changed.

The detailed experimental conditions were as follows:

- Only one food source is placed at a distance of 30 units from the nest.
- An obstacle of length 10 units is placed between the nest and the food source.
- The food source will disappear after a total of 50 times.
- The number of ants in the nest is 100.
- The experiment is repeated 5 times for each condition.

Figure 3.11 shows the simulation environments. Foraging pheromones are shown in green color and exploration pheromone in blue. Figure 3.12 shows the statistics of the experimental results for food collection rate, number of ant deaths, time to finish the food search, and time to find the food source for the first time. They are shown in box-and-whisker plot. The horizontal axis is a comparison under different conditions, where X+ (X–) indicates with (without) an obstacle under condition X.

First, we compare conditions A and B. Under condition B, ants performed better in terms of food recovery rate, number of ant deaths, and search termination time. Let us consider the reason for this. Under condition B, ants that have found the food have an additional function to return to the nest using the exploration pheromone as a guidepost. On the other hand, under condition A, ants must conduct a stochastic search in order to return to the nest after finding the food. This increases the number of ants wandering with the food, resulting in an increase in mortality and a decrease in food recovery rate. In addition, we observed scattered behaviors in which ants were unable to return to their nests with food and deposited pheromones were tracked by other ants, causing them to get lost in groups. This phenomenon is another cause of decreased collection efficiency and increased mortality. There was no predominant difference in the average time until the first discovery of the food source under either condition A or B. This may be because the behavior of ants is practically the same under both conditions until they find the food source.

Now consider a comparison of conditions B and C. Under condition C, in addition to using the foraging pheromones as a guidepost when returning to the nest, the ants in search also act by repelling the exploration pheromones. Figure 3.12(d) shows that the search under condition C is shorter with respect to the termination time for the food, suggesting that avoidance is superior in terms of food supply efficiency. Under condition B, searching ants do not evade the exploration pheromone, which means that ants can search a wide area in a short time. However, ants are unable to distinguish between searched and unsearched areas, resulting in an uneven search area. On the other hand, under condition C, searching ants avoid already searched areas, so the search can be conducted in a uniform manner. But if the ants are surrounded by the exploration pheromone secreted by the preceding ants, they are unable to move. Thus, it is likely under condition B ants tend to deposit exploration pheromones more extensively on average until they find food for the first time. If exploration pheromones are widely distributed, ants may be misled by them. Those exploration pheromones have nothing to do with the direction of return so that the ants may take a detour to return home when they find food and return to their nests. Furthermore, when the ants leave the nest to search for food again, they are forced to follow the detour, resulting in a decrease in search efficiency.

The robustness of the search was further investigated by the placement of obstacles. In terms of ant mortality, obstacles significantly increased the death

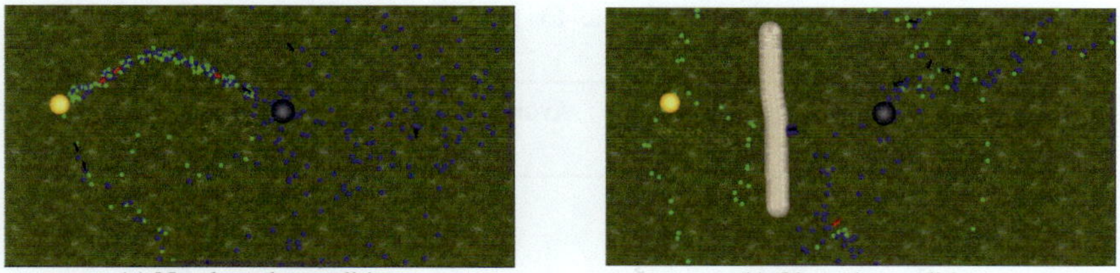

(a) No obstacle condition (b) Obstacle condition

**Figure 3.11:** Simulation of ant search with multiple types of pheromones.

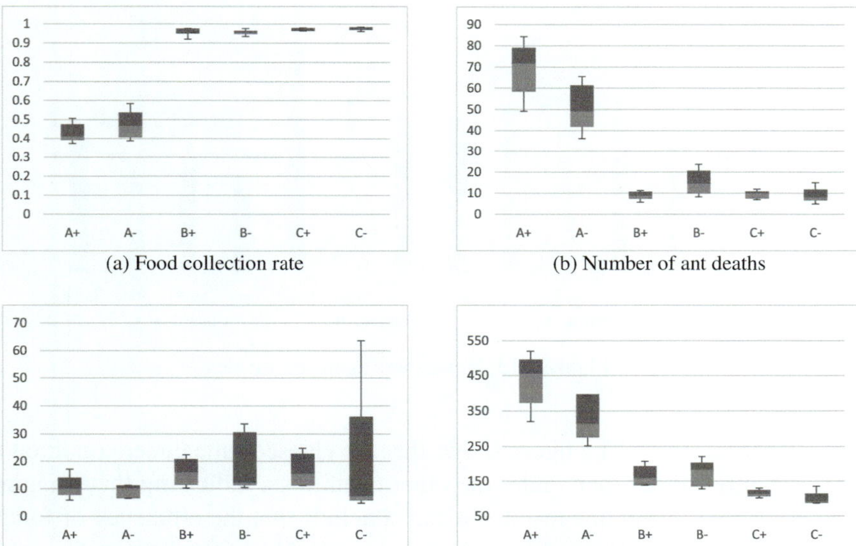


(a) Food collection rate (b) Number of ant deaths

(c) Time to find food (d) Time to end search

**Figure 3.12:** Comparison of ant behavior under different conditions.

counts under condition A, while they decreased under condition B. For the search termination time, it was also significantly worse under condition A, but no such trend was observed under condition B. Under both conditions B and C, the placement of obstacles did not significantly worsen performance. Based on the above results, conditions B and C are considered to be effective in terms of robustness with respect to obstacles.

Robustness to obstacle placement was tested under three conditions (see Table 3.4). For obstacles, four types of environments were exmamined: none, short (20 pixels), intermediate (30 pixels), and long (40 pixels). In the environment with obstacles, they are placed to collide at right angles in the middle of the path after the pheromone trail has been established.

The results are shown in Fig. 3.13. Here we compared the time steps until the foraging was completed for 12 different ways (i.e., three conditions and

**Table 3.4:** Experimental conditions for robustness against obstacle placement.

| Condition | Attenuation ratio $\lambda_E$ | $\lambda_F$ | Avoidance behavior | Difference in evaporation rate |
|---|---|---|---|---|
| A | 0.96 | 0.96 | × | × |
| B | 0.96 | 0.93 | × | ○ |
| C | 0.96 | 0.93 | ○ | ○ |

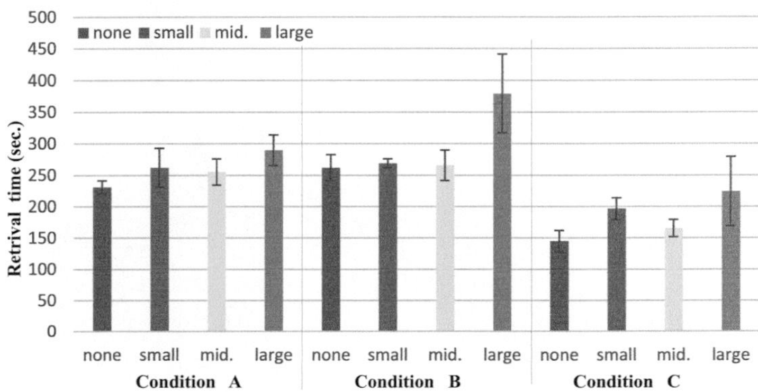


**Figure 3.13:** Robustness against obstacles.

four environments). The figure shows that the elapsed time under condition C is significantly shorter than under the other conditions, indicating that foraging is more efficient. From this result, we can conclude that the efficiency of foraging is improved by avoidance behavior. On the other hand, looking at the process of path formation, under condition C, the shape of the path in the early stages was as random as uder the other conditions, but as time passed, it approached a linear path. In some situations, the effect of different pheromone attenuation rates was not observed. In the previous study [37], the terrain over which the ants could move was limited to a Y shape, which greatly restricted the paths that could be formed. In other words, when the exploration pheromones are on one side of a branch, ants tend to move to the other. On the other hand, as described in this section, in unrestricted terrain, the exploration pheromones become ineffective if they deviate slightly from the path. Simulations that take these effects into account and their validation in actual ants are subjects of future research of interest.

## 3.4 Ants are really diligent?

As the fable at the beginning of this chapter famously states, ants are a symbol of diligence. However, it is known that not all ants in the real world are active, but a certain percentage of ants do not work. In that sense, they are really lazy. Some researchers believe that this allows them to cope more robustly when the environment changes or when some of the ants are lost by accident. In the following, we will experimentally test this hypothesis.

In the simulation described so far, all ants have continued to move in search of food. In reality, there are individual differences among ants; some ants are diligent and keep moving, while others are lazy and do not work until there is no food available to some extent. To achieve this in the simulation, an threshold of "diligence" (an integer value from 1 to 100) is introduced for each individual.

We measure foraging efficiency, which indicates the efficiency of food collection for the entire ant colony. This efficiency index is defined as follows:

- The initial value is 0.
- Increases by one if a worker ant finds food.
- Increases by one if a worker ant brings a food back to the nest.
- Decreases by a constant value (decreasing range $d$) per timestep.

In addition to the exploration and homing states, ants are assumed to enter a lazy (resting) state.

In the lazy state, ants slow down and stop searching for food. Each ant enters the lazy state when the colony's efficiency variable exceeds his diligence threshold. Naturally, the number of ants in the lazy state fluctuates as the value of the efficiency index changes, for example, as a food is found.

Worker ants also rest after a certain level of fatigue. For this purpose, each ant is randomly assigned a value of maximum fatigue between 0.5 and 1.0. The level of fatigue for each individual is calculated based on the distance travelled so far. The ant enters a resting state when it reaches maximum fatigue. The individual that enters the resting state stops and recovers from fatigue until a certain period of time (5 time steps) has elapsed. After a certain period of time has elapsed, the fatigue level is reset and the ant enters the search state again.

In the following simulations, the number of ants is set to 100. The food source is placed in the upper right corner and the nest is in the center. The length $b$ of an obstacle is assumed to be 80 in a standard case (see Fig. 3.14). The search and homing states are together referred to as the diligent state.

First, let us observe how ants behave without the ant fatigue level. This is a simulation with no resting state. Experiments were conducted by setting the decreasing range $d$ (of the efficiency index) as 1, 5, 7, 9 and 10. The time transition of the lazy state is shown in Fig. 3.15(a). The larger the decreasing range, the smaller the number of lazy individuals, and the number of lazy ants is almost constant after a certain time.

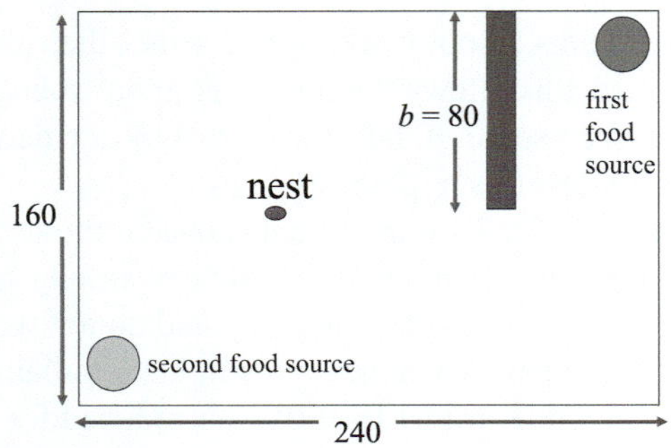


**Figure 3.14:** Experimental environment. Layout of a nest, food sources, and an obstacle.

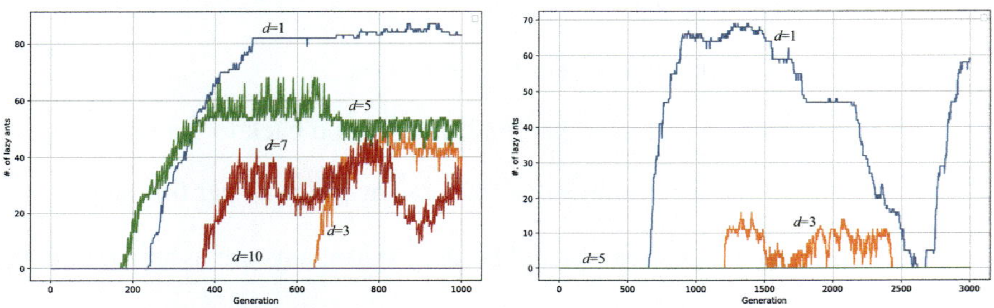


(a) Standard obstacle

(b) Long obstacle

**Figure 3.15:** Experimental results in the absence of resting state.

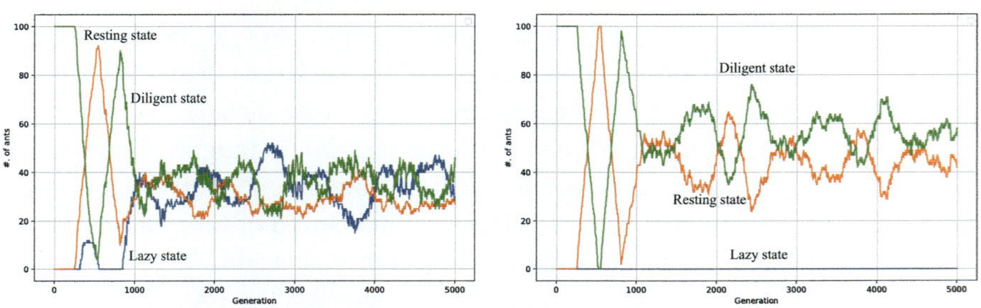


(a) Without lazy ants

(b) With lazy ants

**Figure 3.16:** Comparison with and without resting state.

Robustness was further examined by lengthening the obstacle as $b = 150$. The results are shown in Fig. 3.15(b). When the decreasing range was 5, no individuals became lazy. The behaviour became unstable for the range of 1 or 3,

Next, let us look at a simulation that takes into account the resting state. Figure 3.16 shows the results for the case where the lazy state is present (a) and the case where the lazy state is absent (b), respectively. In both cases, the population of diligent ants is high shortly after the start of the simulation, after which the ants that were moving in search of food become dormant. Note that all ants continue to move industriously until they reach the resting state in case of (a).

In the presence of the lazy state, after a certain period of time, when there is a pheromone trail between the food source and the nest, the population of lazy ants increases and the number of lazy, diligent and fatigued ants is around 30 to 40, which is stable compared to the case without the lazy state. In the absence of the lazy state, the number of diligent and fatigued ants fluctuates over time by approximately 50 individuals each. However, the range of the number of individuals in each state is large, ranging from 30 to 80, and is not stable.

Further simulations were carried out for situations with obstacles of different sizes. Here, the length $b$ of the obstacle was changed as 80, 100, 120 and 140. $b =$ 80 is the original (standard) arrangement. Figure 3.17(a) shows the population of ants in the diligent state when the lazy state is present. The population in the diligent state increased in all cases until they found food. In other words, there were no individuals in the lazy state. On the other hand, after 1000 time steps, when the food was found, the number of diligent individuals decreased and the number of lazy individuals increased as the obstacles were smaller and the food was easier to reach. Conversely, the more difficult it was to find food due to large obstacles, the more diligent individuals increased. However, when the obstacle length $b$ exceeded 120, no lazy individuals were found and all individuals were either dormant or diligent (see Fig. 3.17(b)).

Next, the robustness was tested by adding another food source. The experiment was performed as follows (see Fig. 3.14):

1. At timestep $t = 0$, a food source is placed behind the upper right obstacle.
2. At timestep $t = 1000$, another food source is placed near the lower left nest.

The results of the experiment showed that in the presence of the lazy state, the ants approached the second food source as soon as it was placed. Eventually, the path leading to the first food source behind the obstacle disappeared, and the ants headed only for the second source closer to the nest. In the absence of lazy individuals, ants continued to congregate on both sources after the second food was placed, and they continued to congregate on the source behind the obstacle. Figure 3.18 shows the number of diligent ants. The number of diligent ants was similar in both cases until the second source was placed. However, around 500 timesteps, the number of diligent ants decreased due to the simultaneous resting of those diligent ants. After the second food source was placed, the number of diligent ants decreased in the presence of the lazy state and settled at about 20 individuals. On the other hand, in the absence of the lazy state, the population oscillated as a large number of individuals went dormant at the same time.

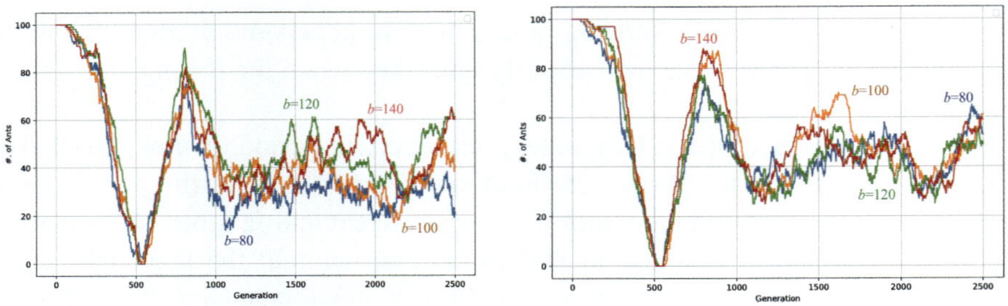


(a) With lazy ants (b) Without lazy ants

**Figure 3.17:** Comparative experiments with different lengths of obstacles.

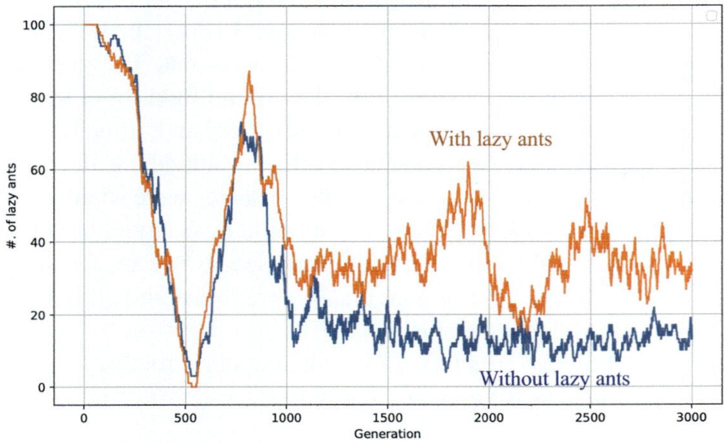


**Figure 3.18:** The number of diligent ants with generations.

To summarize the results of the above experiments, the transition between the diligent, dormant, and lazy states is as follows:

- The number of individuals in the resting state increases as the number of individuals who worked diligently exceeds a certain amount of movement and reaches the limit of fatigue.
- As the number of individuals in the dormant state increases, the colony's efficiency index decreases, and individuals who had been in the lazy state begin to work diligently. As a result, the number of individuals in the lazy state decreases slightly behind the above state, and the number of individuals in the diligent state increases.
- The individuals that had been in the lazy state take a sufficient rest and start working again in the diligent state. The colony's efficiency index

increases accordingly, and the individuals with a larger threshold become lazy state.

Thus, the following advantages are possible with the lazy state:

1. The number of diligent individuals is kept constant.
2. The number of individuals that become dormant can be kept at a low level.
3. The overall population can respond to the loss of a group of hardworking individuals.

There is a well-known law of ecology called "Worker Ant Theory." This law states that among worker ants, 20% are diligent individuals who work continuously, 60% work or slack off depending on the situation, and the remaining 20% are lazy all the time. Based on the results of previous experiments, ants can be classified into the following three types:

- ants that work diligently are individuals with a large threshold.
- ants that change depending on the situation are individuals with a medium threshold.
- ants that are lazy all the time correspond to individuals with a small threshold.

Furthermore, we have confirmd that the presence of a lazy state has the following advantages:

- Can stabilize the colony in terms of the amount of food and the number of worker ants, etc.
- Can be flexible to dynamic environments.
- Can improve energy efficiency.

## 3.5 How do ants build bridges?

Altruism refers to behavior that prioritizes benefits to others rather than self and sometimes involves acts of self-sacrifice in order to aid others. Some army ants construct living bridges with their own bodies when they find holes or gullies as obstacles to their marching routes as shown in [4] (see Fig. 3.19). Such philanthropic acts are different from the regular behavior of the ants, e.g., foraging for and transport of food. However, if more ants participate in bridge construction than that is required or if they construct bridges at sites where those are unnecessary, they may actually hamper the food gathering performance of the whole colony. But, in nature, the ants are very keen to balance these actions as per requirement and it has been confirmed that because of such altruistic activity the

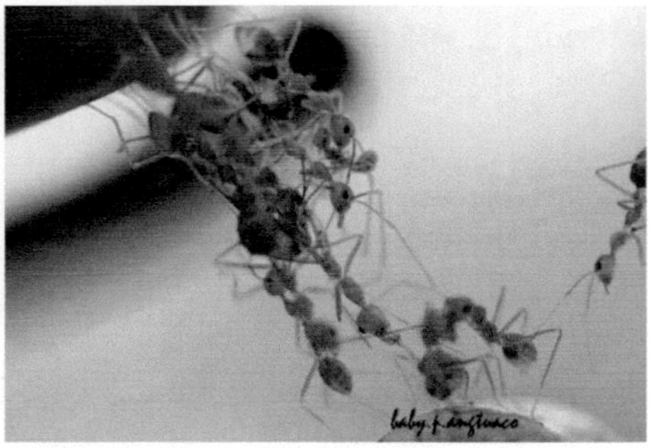

**Figure 3.19:** A scene of building a living bridge by army ants (photo courtesy of Prof. Salvacion P. Angtuaco [2]).

performance is improved for the group as a whole. In an experiment by Powell et al., it was found that the foraging capacity of the army ant colony increased by up to 26% due to this altruistic behavior [101]. In this section, this altruism of ants is modeled and examined in a multi-agent simulation environment.

### 3.5.1 Defining the problem

The present simulation serves as a model for the foraging behavior and the altruism of ants. The actions include foraging for and transport of food and communications with neighboring ants using pheromone. The nest is the starting point of the ants and also the point to which the ants return with food (see Fig. 3.20). The pheromone is released by an ant when it finds food. Just as in nature, once secreted the pheromone attenuates and disperses, thus disseminating information among the ants about the food locations. A gully hinders movement of ants and fundamentally prevents the ants from passing over it. However, if an ant shows altruism and forms a living bridge over the gully, other ants can pass over the

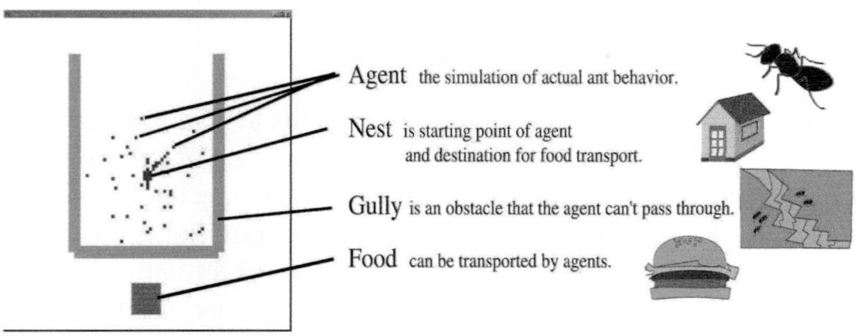


**Figure 3.20:** Simulation environment.

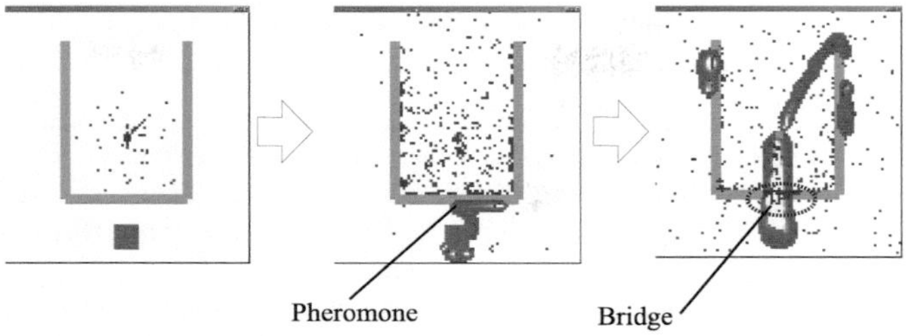


**Figure 3.21:** Simulation of bridge generation by army ants.

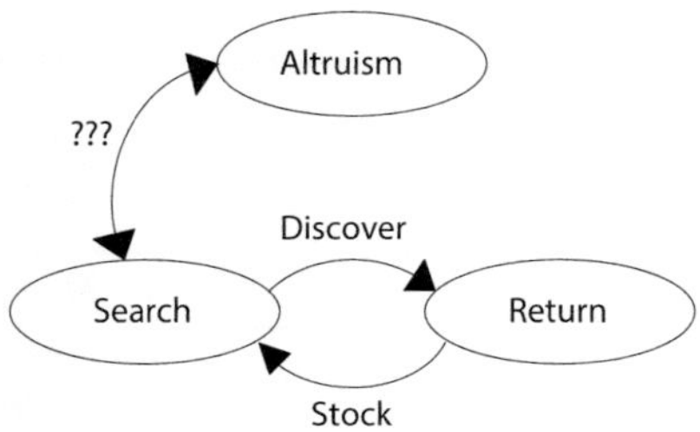


**Figure 3.22:** State transition of ants.

gully (see Fig. 3.21). The ants move in accordance with the state transition diagram shown in Fig. 3.22. The behavior of ants in different states are shown in Table 3.5.

The problem is to determine the conditions that induce the transition to the Altruism state. But it is not concretely known how ants decide the site and timing of living-bridge construction and when they cease the bridge formation. Therefore, in this section, several hypotheses are proposed as the altruism initiation conditions and experiments are performed for verification.

### 3.5.2 *Judgment criteria for entering altruism state*

Two hypotheses have been proposed as the judgement criteria for altruistic activity by army ants.

**Model 1:** Based on the presence of neighboring ants

An ant will start the formation of a living bridge over a gully only when neighboring ants are present. Hypothetically, this approach will be more efficient compared to forming a bridge blindly because when there are neighboring ants the probability is high that they will utilize the shortcut.

**Table 3.5:** States and behaviors of ants.

| State | Behavior |
|---|---|
| Search | This is the initial condition of the ant and it continues random work until food is found. When food is found, there is a transition to the Return state. Transition to Altruism state is also possible under "certain" conditions. When pheromone is sensed, the ants are drawn to the higher concentrations. |
| Return | The food is returned to the nest. In this state the ant moves toward nest secreting pheromone. After reaching the nest, the ant transits to the Search state. An ant in Return state knows the position of the nest. |
| Altruism | A bridge is constructed across the gully. While in this state, movement is impossible for an ant. When certain conditions are met, the bridge is abandoned and the ants transit to the Search state. |

**Table 3.6:** Properties used in two types of models.

| | Model 1 | Model 2 |
|---|---|---|
| # of steps | 700 | 700 |
| time | 10 | 150 |
| radius | 2 | - |
| pheromone threshhold | - | 30 |

**Model 2:** Based on the presence of pheromone
As described earlier, ants secrete pheromone when they find food, and this pheromone is used to disseminate information among ants about the location of the food source. The places where pheromone concentrations are higher than a fixed level are the locations that many ants have passed and/or will pass through in the future. Therefore, a living bridge can be formed judging the pheromone concentration.

In both models, ants leave the bridge after a fixed amount of time passes. We actually used fixed properties optimized by genetic algorithms (shown in Table 3.6, [58]). In order to judge their validity, these hypotheses were fed into the simulation and their usefulness was verified empirically.

The two scenarios shown in Fig. 3.23 were used in the experiment. In these experiments, performance was measured using the number of food items collected within a fixed period of time. Each experiment was repeated 10 times with 20 to 180 ants, increased by 20 at a time, and the mean values were compared.

The experimental results from the simple map are shown in Fig. 3.24. The number of ants is shown along the horizontal axis and the number of food items collected within a fixed amount of time is shown along the vertical axis. In the simple map, **Model 1** showed slightly higher performance, but the differences were small and almost no difference in overall efficiency was observed.

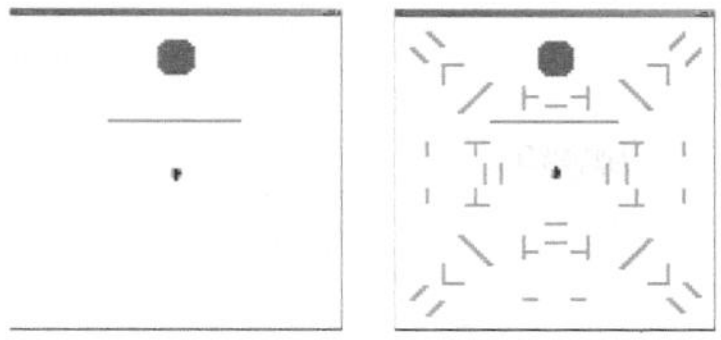


**Figure 3.23:** Maps for experiment.

Experimental results using the difficult map are shown in Fig. 3.25. On the whole, **Model 2** performed better in the difficult map. Figure 3.25(b) shows experimental observations for the difficult map on a different scale. The horizontal axis represents the number of ants, whereas the vertical axis represents the ratio of the total number of time steps ants crossed bridges to the total number of time steps ants helped to form bridges. This ratio indicates how useful the formed bridges were. From the data, it was found that **Model 2** yielded higher values than **Model 1**. For **Model 1**, the ratio was usually about one. This means that even though a bridge was formed, neighboring ants would have not used it efficiently. This was because in the difficult map, unlike the simple map, gullies were present at various locations causing bridges to be formed at unnecessary sites with **Model 1**. With **Model 2** higher ratios were found compared to that found with **Model 1**. Although it is not evident from the graph, in **Model 2** the bridges were formed only at those sites that were necessary for bringing food to the nest. This was because the pheromone was secreted along the way from the food-source to the nest. The concentration of pheromone indicated the optimal sites for bridge construction. Hence, both the timing and sites of bridge construction were superior in **Model 2**. However, **Model 2** suffers from the drawback that bridges cannot be formed until the foraging sites have been found. In nature, cases are also observed where bridges are formed at necessary sites before foraging sites are found. Thus, for altruistic activity like bridge formation, ants may use the pheromone method along with some other judgement criteria such as the one stated in **Model 1**.

### *3.5.3 Judgment criteria with reference to chain formation*

Chain formation is another philanthropic cooperative behavior similar to bridge formation. Chains in this case refer to structures formed by the bodies of the ants when the ants encounter extreme differences in heights during their marches. In this way, it is possible for other ants to move safely from one height to another. In their research, Lioni et al. [79] observed the chain formation behavior of ants in nests installed in the laboratory. The results showed that the probability of participation in chain formation $P_e$ and probability of abandoning chain formation

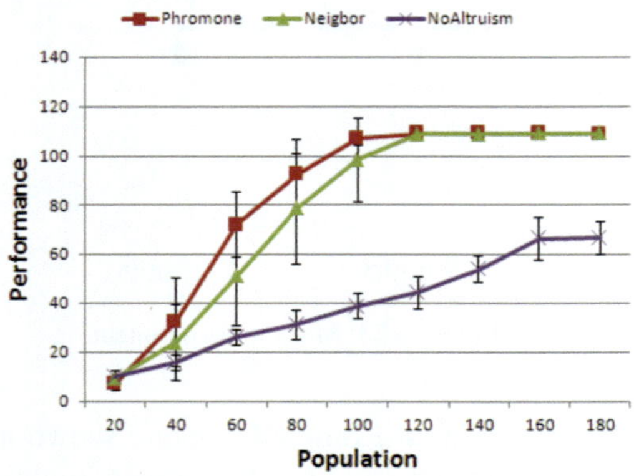


**Figure 3.24:** Simple map – Experimental results.

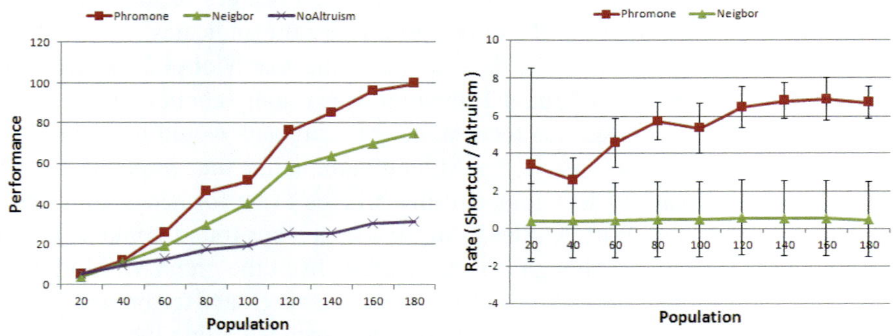


(a) Result 1

(b) Result 2

**Figure 3.25:** Experimental results for difficult map.

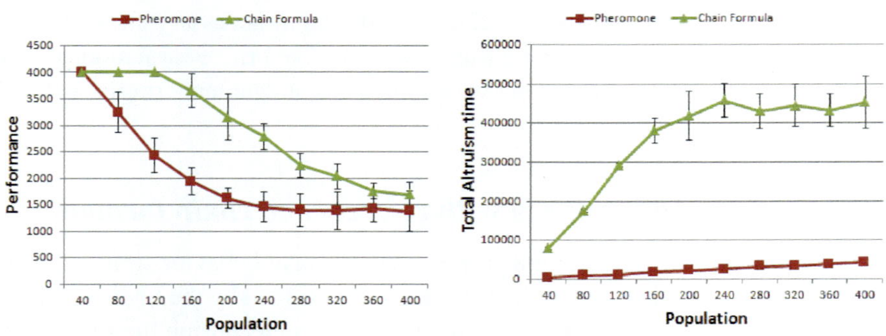


(a) Foraging time

(b) Altruistic activity

**Figure 3.26:** Performance comparison (chain formation).

$P_s$ can be approximated by the following equations:

$$P_e = C_{e0} + \frac{C_{e1}X}{C_{e2} + X}, \quad (3.4)$$

$$P_s = C_{s0} + \frac{C_{s1}X}{C_{s2} + X^{\nu}}, \quad (3.5)$$

where $X$ is the number of ants participating in chain formation. According to these equations, if many ants are contained in the formed chain, then it is easier for them to participate in chain formation but more difficult for them to stop.

To justify the proposed model of pheromone concentration as the criteria for transition to altruism state, a comparative study was performed with the Lioni's model of chain formation.

In Fig. 3.26(a), the number of ants is shown on the horizontal axis and the time until the completion of foraging on the vertical axis. It was found that for some population sizes when pheromone concentration is used as the judgment criteria, foraging takes shorter time than that is required for chain formula.

In Fig. 3.26(b), the number of ants is shown on the horizontal axis and the cumulative time during which the ants are engaged in altruistic behavior on the vertical axis. It was also observed that when pheromone concentration is used as judgment criteria, the total time during which the ants are engaged in altruistic activity is shorter and is less affected by the population size. On the other hand, when the formulas of Lioni et al. (i.e., eqs. (3.4) and (3.5)) are applied, the time engaged in altruistic behavior increases with the number of ants. Figure 3.27 compares another aspect of the models. When pheromone concentration was used as judgment criteria, bridges were constructed at the required ones. However, when the formulas of Lioni et al. were applied, bridges were constructed at many sites other than the required sites. It is also clear from this figure that with the Lioni et al. model, fewer ants are in search state as many of them are in altruism state.

Procedures using eq. (3.4) of Lioni et al. featured a higher probability of altruistic behavior at sites where ants are apt to congregate. Therefore, more altruistic behavior is expected to occur close to the foraging site and the nest or in between these sites. In Lioni's model, the altruistic behavior is possible without finding foraging sites and this is an advantage over the proposed model based on pheromone concentrations. Nevertheless, the simulation results showed that in terms of performance, measured as foraging speed, the proposed model was superior to Lioni's model. The possible reason behind this could be that in Lioni et al. experiment calculations were performed by limiting the chain formation sites to one, hence, their model could not be directly applied to an environment with a series of bridge formation sites as used here. Therefore, in consideration of biology, etc. of army ants, we need to combine the pheromone concentration based model with other judgment criteria.

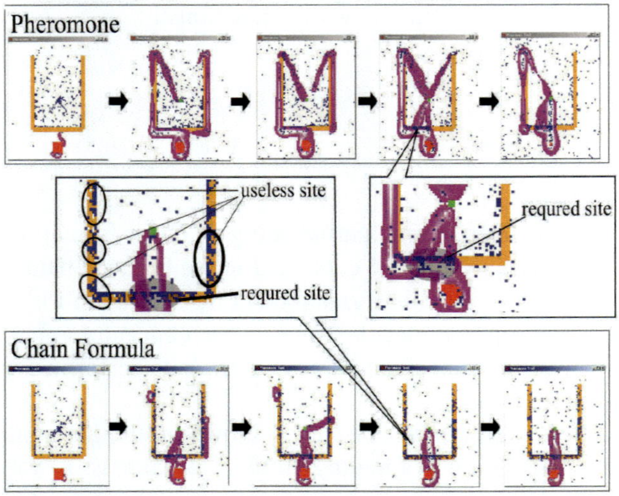


**Figure 3.27:** Comparison of bridge construction sites.

### *3.5.4 Changes in strategy based on number of ants*

It has been confirmed that the group behavior of army ants is seriously affected by the number of ants that are active [79, 80]. For example, when few ants are available for chain formation, chains are not formed. When a large number of ants are available, chains are formed at several sites. On the other hand, when the number of active ants is moderate, initially several chains are formed. After a certain time, extension of most of the chains stops and the chains gradually decrease in size and eventually the extension of only one chain continues. However, it is still not clear how the ants count the number of neighboring ants and how this number affects their behavior.

In order to monitor the effect of group size on the activity of ants, we performed experiments using eqs. (3.4) and (3.5) extended with a minimum limit on group size as an additional condition of chain formation. We compared this scheme with the one that does not take into account the group size. The experiment was performed using two maps shown in Fig. 3.28. The results of the experiments are shown in Fig. 3.29. The horizontal axis shows the number of ants, and the vertical axis shows the performance in terms of the number of food items collected within a fixed time. In these figures, "with Check Neighbor" represents the procedure taking the number of neighboring ants into consideration and "without Check Neighbor" indicates the procedure not taking the number of neighboring ants into consideration.

In Map 1, the method that did not take into account information about neighboring ants showed high performance. This was because the conditions for bridge formation were relaxed and hence bridges could be formed at an early stage and food can be found easily.

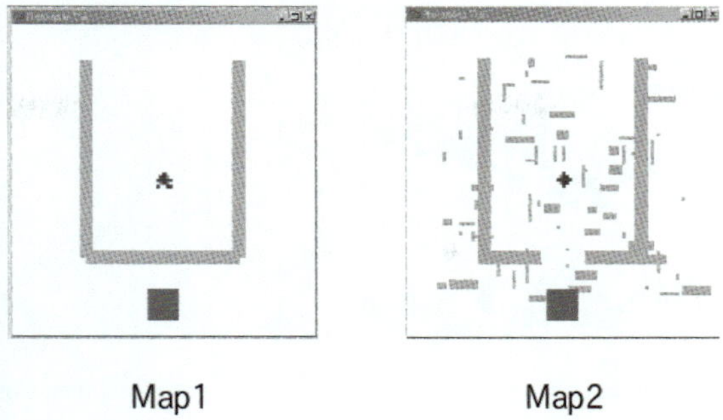


**Figure 3.28:** Maps used to study the effect of number of ants.

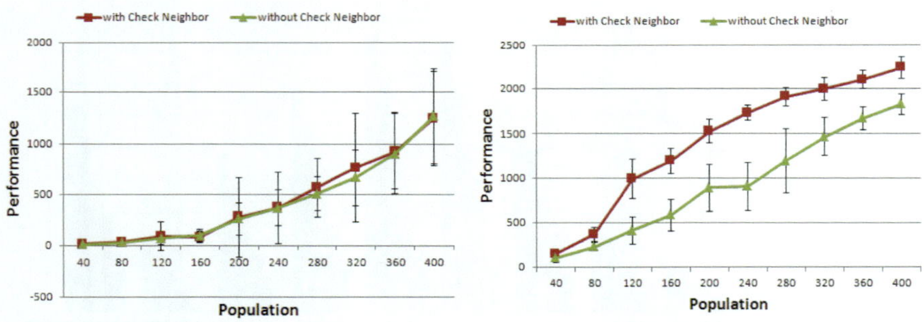


(a) Map 1 (b) Map 2

**Figure 3.29:** Effect of neighborhood knowledge.

Map 2 was used to investigate whether intelligent behavior can be achieved by avoiding unnecessary bridge formation where a shortcut is not especially necessary for food collection. In this case, better results were obtained with the method that checks the number of neighboring ants.

Figure 3.30(a) shows how the bridges extend in size with time for Map 1. In the figure, "1st" refers to the largest bridge at the time and "2nd" refers to the next largest bridge. The horizontal axis shows time steps and the vertical axis shows the two largest bridges. It is apparent from the graph that at first, several bridges coexist and extend for about the same length, but finally the differences become greater. Figure 3.30(b) shows the data obtained in a biological experiment in the research of Lioni et al. [80]. When chains were formed at two sites, records were kept on how each of the chains extended. In the figure, "1st" and "2nd" show the sizes of the chains at each site.

As can be seen from the two figures, these curves, i.e., simulation data and data from the biological experiment, look quite similar.

From the previous experiments, it seems that our model has many properties similar to actual army ant behavior. To emphasize the similarity between the

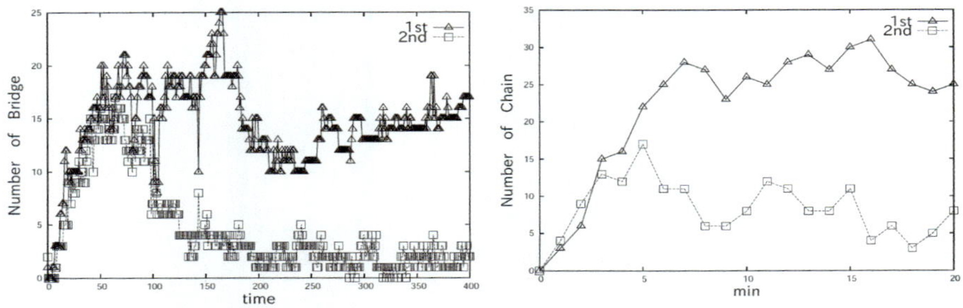


(a) Simulation data (b) Real data from [80]

**Figure 3.30:** Changes in size of bridge.

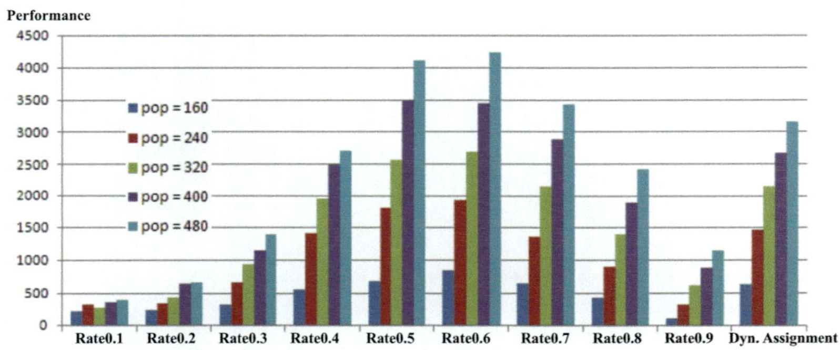


**Figure 3.31:** Experimental results with task assignment.

simulator ants and the actual army ants, it is important to compare experimental data.

As a first step, the experiment was performed using a simulator that has ants with a fixed task assigned. Task assignment is one of the signatures that is observed in army ants. Army ants have tasks that depend on someone's rank. Here we consider the following two different roles for ants in our simulator:

**role A:** Search and transport food.

**role B:** Build a bridge to support role A.

We carried out the experiment by assigning ants to these two roles with different ratios. Figure 3.31 shows the experimental results where the performance was compared in terms of the number of food items collected within a fixed time. Rate 0.1 means that 10% of the ants were assigned to role B in the simulation. "Dynamic Assignment" labels the experimental results obtained by the simulator used in Section 3.5.4 where the ants have no fixed role.

Results in Fig. 3.31 indicate that a fixed division of roles may be better than a dynamic one. Especially Rate 0.5 and Rate 0.6 are better than other ratios.

Although "Dynamic Assignment" was not the best, it performed competitively on an average.

### 3.5.5 How much profit do ants gain from building the bridge?

In this section, Reid et al. [108] conducted experiments based on actual ants to determine when to build (or not to build) a bridge for terrain shown in Fig. 3.32(a). In the following, we will run simulations and compare them to the measured data.

In this experiment, the field consists of four bridges of length $L0$ and width $w_A$ joined together. Let $\theta$ be the angle formed by the two bridges in the middle. In the following simulation, we set as $L0 = 80$ and $w_A = 24$. The ants are only allowed to move on these bridges. The simulation environment is shown in Fig. 3.32(b). In this figure, the white area is the bridge. A nest (black circle) is positioned at the left end of this bridge and a food source (orange circle) at the right end. New ants are generated from the nest at regular intervals and search the field at random. They return to the nest when they find food.

The ant transitions between the following four states:

**Random search state** The ants search the field at random. They also release food pheromone at this time. Ants are in this state immediately after they emerge from the nest.

**Pheromone search state** The ant determines its direction of travel based on the number and direction of nearby food pheromones, rather than randomly. Ants in the random search state transition to this state when they find a food pheromone.

**Homing state** If homing pheromones are present nearby, the ant determines its direction of travel based on the

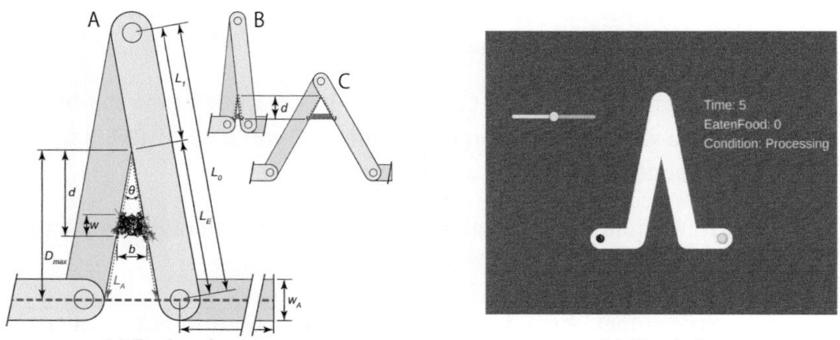


(a) Real environment (b) Simulation

**Figure 3.32:** Experimental environment for ants to build bridges.

number and direction of homing pheromones. If not present, the direction is determined randomly. When ants find food, they transition to this state and bring the food back to the nest and release homing pheromone.

**Bridge state** The ant stops moving and becomes part of a bridge. Other ants can move over ants in the bridge state. Ants transition to this state if certain conditions (see below) are met in the random or pheromone search state.

The search performance of ants in creating a bridge state is compared under the following three conditions:

**No altruistic** Ants do not form a bridge.

**#. Ant** Ants transition to a bridge state if there are more than a certain number ($N_{TH}$) of ants in the surroundings.

**#. Pheromone** Ants transition to a bridge state when the total amount of homing and feeding pheromones in the surroundings exceeds a certain value ($P_{TH}$).

The parameters in the simulation are as follows:

- Internal activity $E$: Decreases at regular intervals; if $E$ falls below a certain threshold ($E_{TH}$ ), the ant dies.
- Pheromone threshold $P_{TH}$: Under **#. Pheromone** conditions, ants transition to a bridge state when the number of surrounding pheromones exceeds this value. In this experiment $P_{TH} = 8$.
- Ant threshold $A_{TH}$: Under **#. Ant** conditions, ants transition to a bridge state when the number of ants present in the surroundings exceeds this threshold. In this experiment, $A_{TH} = 1$.
- Time threshold $T_{TH}$: After this time has elapsed since the bridge state, the state is terminated and the random search state is entered. In this experiment, $T_{TH} = 70$.
- Ant makespan $M$: Interval at which ants are generated. We set $M = 1.5$ in this experiment.

In this experiment, five simulations of three conditions (**No altruistic**, **#. Ant**, **#. Pheromone**) are performed for 14 different field angles, i.e., $\theta = 16, 18, 20, 21.15, 22.5, 25, 27.5, 30, 32.5, 35, 40, 50, 60, 70$. The average number

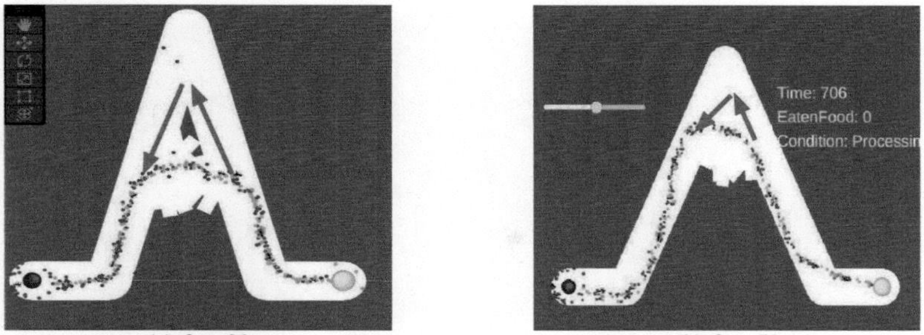


(a) $\theta = 20$ (b) $\theta = 50$

**Figure 3.33:** Examples of bridges built by ants.

of ants bringing food back to the nest at elapsed time = 2000 is compared. Also, we observe at what $\theta$ ants build bridges.

The simulation results showed that the ants made bridges at all angles. Figure 3.33 shows the ants moving in the simulation for $\theta = 20$ and $\theta = 50$. The small blue and green dots are homing and food pheromones, respectively. The trail connecting them corresponds to the path taken by the ants. The red arrows in the figure are the shortest paths that the ants would take to reach the nest if they did not build bridges.

Figure 3.34(a) shows a plot of angle $\theta$ on the horizontal axis and score on the vertical axis. Under **#. Ant** condition,the score changed little with $\theta$, whereas under the other two conditions, it decreased with increasing $\theta$. When $\theta < 27.5$, higher score results were obtained under **#. Ant** and **#. Pheromone** conditions than **No altruistic** condition. When $\theta$ was between 27.5 and 32.5, there was no difference between the three conditions. When $\theta > 32.5$, a higher score was obtained under **No altruistic** condition. This may be due to the fact that as $\theta$ increases, the travel saving effect of the ant-built bridges decreases. As can be seen from the figure, the travel savings from the bridges built by the ants is higher for $\theta = 20$. Figure 3.34(b) plots the travel savings for different $\theta$. The amount of travel saved is greatest for $\theta$ between 18 and 20, and then the amount of travel saved decreases as $\theta$ increases. It is costly for ants to build bridges. Ants on bridges cannot move and often die as a result. Thus, when $\theta > 32.5$, the benefit of saved mobility is less than the cost of building bridges, and altruistic behavior becomes counterproductive. This phenomenon can be seen in the decrease in the score in Fig. 3.34(a).

Let us compare the results of this simulation with actual ants' behavior [108]. Figure 3.35 is a graph plotting $\theta$ on the horizontal axis and final bridge position on the vertical axis. Final bridge position is the distance between the top of the field and the bridge created by the ants (see Fig. 3.33). In all cases, $\theta$ increases rapidly up to around 20, and then decreases as $\theta$ increases. Note that the simulation results are qualitatively consistent with the actual ant behavior.

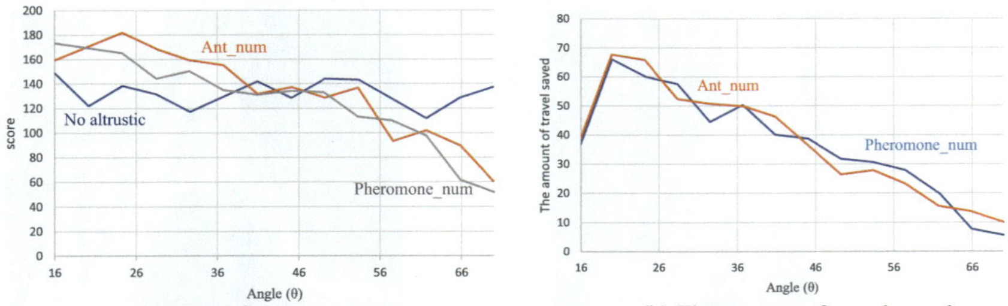


(a) Score

(b) The amount of travel saved

**Figure 3.34:** Experimental results.

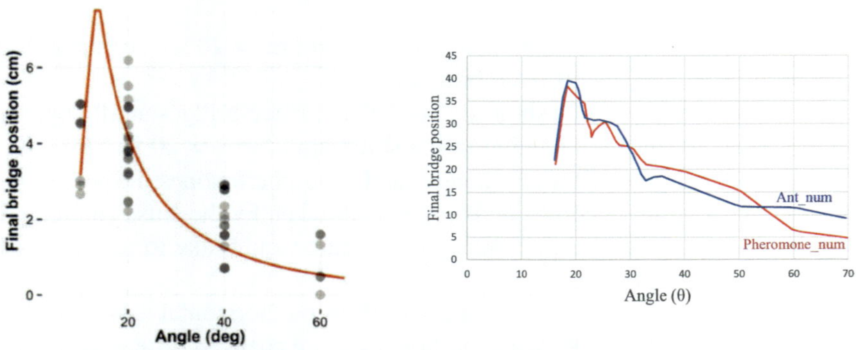


(a) Real data from [108]

(b) Simulation data

**Figure 3.35:** Experimental environment for ants to build bridges.

## 3.6 ACO and its extension

Optimization algorithms based on the collective behavior of ants are called ACO (Ant Colony Optimization) [30].

First, let us consider TSP (traveling salesman problem). In TSP (traveling salesman problem), a set of cities and a travel cost (e.g., distance) between each two cities are given. The aim is to find the cycle path that goes through all cities exactly once and returns to the starting city (this is called a Hamiltonian cycle or path) with the lowest total travel cost (see Fig. 3.36). There is no efficient algorithm that will solve the TSP; in all cases, exhaustive enumeration is required in order to find the optimum solution. Consequently, as the number of cities grows, we see a dramatic leap in the complexity of the problem. This is called a "combinatorial explosion," and is an important issue (an NP-complete problem) in the field of computer science. TSP is applied in areas such as commodity distribution cost and LSI pattern technology.

**Figure 3.36:** Image of a traveling salesman.

ACO using a pheromone trail model for the TSP uses the following algorithm to optimize the travel path:

**Step 1** Ants are placed randomly in each city.

**Step 2** Ants move to the next city. The destination is probabilistically determined based on the information on pheromones and given conditions.

**Step 3** Repeat until all cities are visited.

**Step 4** Ants that make one full cycle secrete pheromones on the route according to the length of the route.

**Step 5** Return to **Step 1** if a satisfactory solution has not been obtained.

ACO algorithm can be outlined as follows. Take $\eta_{ij}$ as the distance between cities $i$ and $j$. In **Step 2**, the probability $p_{ij}^k(t)$ that an ant $k$ in city $i$ will move to city $j$ is determined by the reciprocal of the distance $1/\eta_{ij}$ and the amount of pheromone $\tau_{ij}(t)$ in the following way (see Fig. 3.37):

$$p_{ij}^k(t) = \frac{\tau_{ij}(t) \times \eta_{ij}^\alpha}{\sum_{h \in J_i^k} \tau_{ih}(t) \times \eta_{ih}^\alpha}., \tag{3.6}$$

where $J_i^k$ is the set of all cities that the ant $k$ in city $i$ can move to (has not visited). The condition that ants are more likely to select a route with more pheromone reflects the positive feedback from past searches as well as a heuristic for searching for a shorter path. ACO can thereby include an appropriate amount of knowledge unique to the problem.

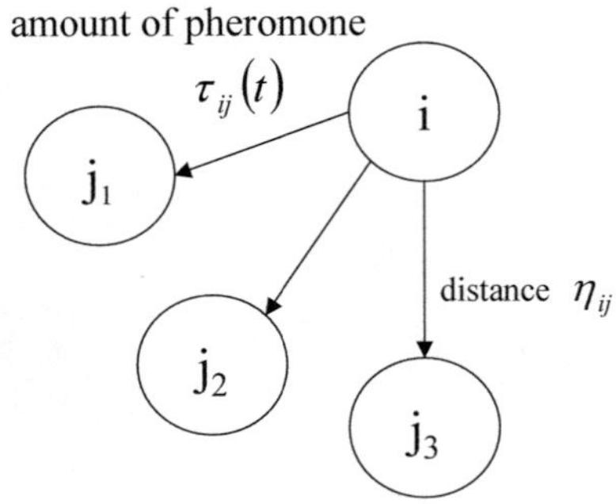


**Figure 3.37:** Path selection rule of ants.

The length and pheromone accumulation of paths between cities are stored in a table. Ants can recognize information on their surroundings, and probabilistically decide the next city to visit. The amount of pheromones added to each path after every cycle is inversely proportional to the length of the cycle path. More formally, the pheromone table is updated by the following equations:

$$Q(k) = \text{the reciprocal of the path that the ant } k \text{ found,} \tag{3.7}$$

$$\Delta\tau_{ij}(t) = \sum_{k \in A_{ij}} Q(k), \tag{3.8}$$

$$\tau_{ij}(t+1) = (1-\rho)\cdot\tau_{ij}(t) + \Delta\tau_{ij}(t), \tag{3.9}$$

where $A_{ij}$ is the set of all ants that moved on a path from city $i$ to city $j$. The amount of pheromone added to each path after one iteration is inversely proportional to the length of the paths that the ants found (eq. (3.7)). The results for all ants that moved through a path are reflected in the path (eq. (3.8)). Negative feedback to avoid local solutions is given as an evaporation coefficient (eq. (3.9)), where the amount of pheromone in the paths, or information from the past, is reduced by a fixed factor ($\rho$).

ACO is an effective method to solve TSP compared to other search strategies. Table 3.7 shows the optimized values for four benchmark problems and various minima found using other methods [30]. The integers in brackets indicate the number of candidates investigated. ACO is more suitable for this problem compared to methods such as genetic algorithm (GA), simulated annealing (SA) and evolutionary programming (EP).[7] However, it is inferior to the Lee-Kernighan method (the current TSP champion code). The characteristics that specialized methods perform better in static problems are shared by many metaheuristics (high-level strategies which guide an underlying heuristic to increase their performance). For more complex problems, such as TSPs with asymmetric distances

[7] A method similar to GP (Genetic Programming) that focuses on evolving finite state machines. But the structure to be optimized is fixed, while its numerical parameters are allowed to evolve.

**Table 3.7:** Comparison between ACO and metaheuristics.

| TSP prob. | ACO | GA | EP | SA | Optimal |
|---|---|---|---|---|---|
| Oliver 30 | 420 | 421 | 420 | 424 | 420 |
| | [830] | [3200] | [40,000] | [24,617] | |
| Eil 50 | 425 | 428 | 426 | 443 | 425 |
| | [1,830] | [25,000] | [100,000] | [68,512] | |
| Eil 75 | 535 | 545 | 542 | 580 | 535 |
| | [3,480] | [80,000] | [325,000] | [173,250] | |
| KroA 100 | 21,282 | 21,761 | N/A | N/A | 21,282 |
| | [4,820] | [103,00] | [N/A] | [N/A] | |

between cities or dynamically changing cities, no established method has yet been developed. Therefore, ACO is one of the most promising research methods.

Figure 3.38 shows an example of a TSP solution method using ACO. (a) is an initial random city configuration (15 cities). This simulation was run by the TSP solver by ACO described in Appendix A.1. In the figure, pheromone trails are indicated by yellow-green lines. The darker the line, the higher the pheromone concentration. The shortest paths found at the moment are shown in red. The search proceeds from (b) to (e) so as to find the optimal solution. Durng the execution of ACO, the city is moved as indicated by the arrow (see Fig. 3.38(f)). Despite this movement of the city, the ACO search continues to find the shortest path in the new configuration (Figs. 3.38(g) and (h)). Note that the new path is a partial reconstruction of the original shortest path. This ability to adapt to dynamic changes in the environment demonstrates the robustness of the ACO search. The problem in Fig. 3.38 is called a dynamic TSP and is used in real-world applications such as delivery planning and traffic jam avoidance for car navigation systems. The strength of ACO as well as PSO and evolutionary computation is diversity in the population. This feature helps different individuals to respond to different changes and allows the population as a whole to adapt to a dynamic environment. This adaptation is not always easy with a single learning machine, such as neural networks or reinforcement learning.

In addition to TSP, ACO can solve combinatorial optimization problems that are attributed to path generation problems. ACO has also been applied to clustering and sorting. In addition, ACO has been applied to a variety of real-world problems including deep neural evolution (see Section 5.1). For example, we have proposed a continual learning method for LSTM (Long Short Term Memory) using ACO to reduce catastrophic forgetting. The method iteratively optimizes the internal structure of the LSTM in parallel with the training of the current task, using the model performance as fitness. It also extends the replay-based method and utilizes two kinds of memory buffers to reduce overfitting to the memory. See [57, Section 6.3] and [68] for details.

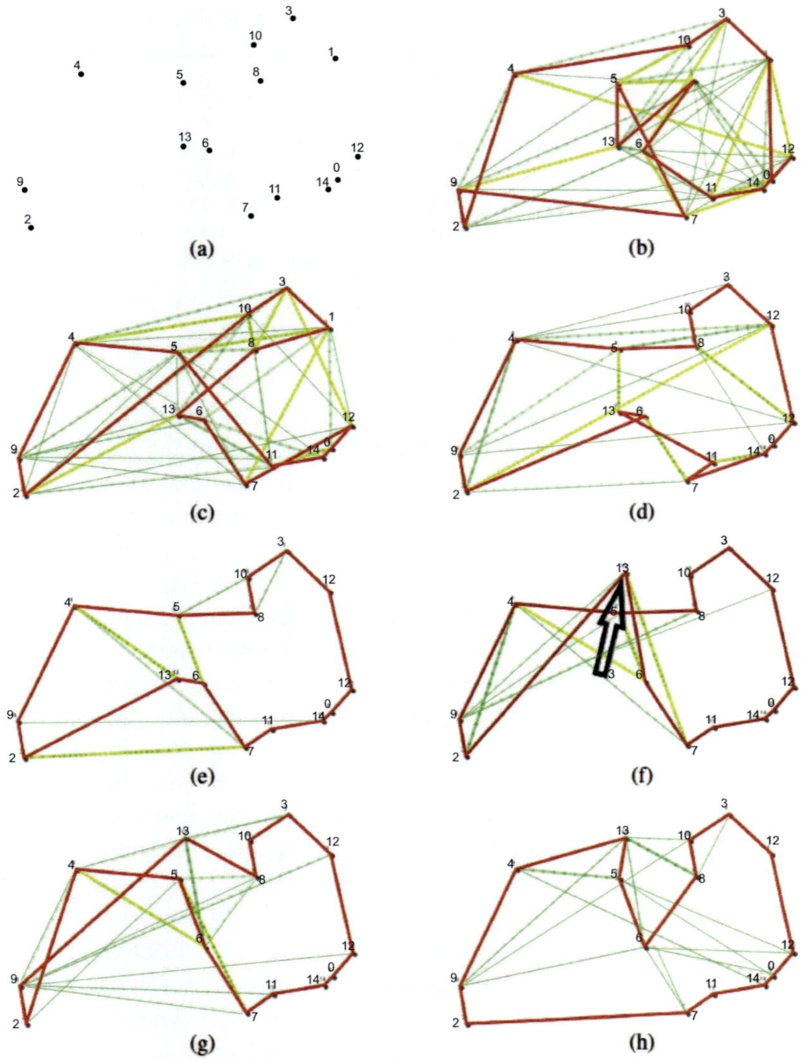


**Figure 3.38:** Dynamic TSP.

Population-based ACO (P-ACO) has been proposed as an extension of ACO [45]. The general flow of P-ACO is the same as ACO. Equation (3.6) is used for search by ants, but the pheromone update method is different. Let $P(t)$ be the set of ants at time step $t$ and $K(t)$ (of size $k$) be the storage memory (see Fig. 3.39). The best ant (i.e., the ant that took the path with the smallest total distance) among $P(t)$ is added to $K(t)$ so as to generate $K(t+1)$. At this time, if the memory already contains $k$ ants, the oldest ant is discarded so that the memory size is always kept at $k$. In other words, $K(t)$ stores the path taken by the best ant from step $t-k$ to $t-1$.

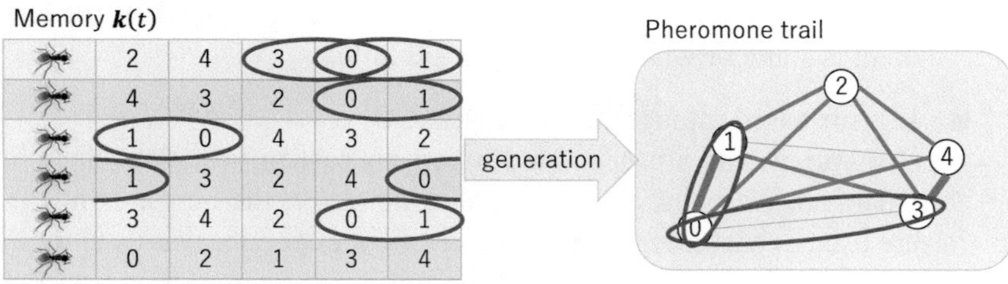


**Figure 3.39:** Population-based ACO.

When the memory $K(t)$ is updated at step $t$, the pheromone is also updated with the following equation:

$$\tau_{i,j}(t+1) = \begin{cases} \tau_{i,j}(t) + \Delta\tau^{a}_{i,j}(t) & \text{When an ant } a \text{ is added to } K(t), \\ \tau_{i,j}(t) - \Delta\tau^{a}_{i,j}(t) & \text{When an ant } a \text{ is removed from } K(t), \end{cases} \tag{3.10}$$

$$\Delta\tau^{a}_{i,j}(t) = \begin{cases} (\tau_{max}(t) - \tau_0)/k & (i,j) \in T^{a}, \\ 0 & (i,j) \notin T^{a}, \end{cases} \tag{3.11}$$

where $\tau_{max}$ and $\tau_0$ are the upper and initial pheromone values between each city, respectively. $T^a$ is the Hamiltonian cycle that ant $a(\in P(t))$ went through. $\tau_0$ is set to be $1/(n-1)$ when the number of cities is $n$.

In P-ACO, unlike ACO, past information is stored in memory with a fixed size limit. Furthermore, the upper and lower limits of the pheromone are fixed. Since the data in the memory is lost in order of oldest to newest, it is possible to track new optimal solutions even after the environment has changed, as in a dynamic TSP, without getting too caught up in old information.

An improved method of P-ACO is immigrant schemes [84, 85]. When generating $K(t+1)$, the best ants of $P(t)$ are added to $K(t)$, and then some of the memory is swapped with the transferred ants to increase the diversity of $P(t)$ and make it easier to adapt to dynamic environments. In immigrant schemes, the short-term memory $K_{\text{short}}(t)$ (size $k_s$) corresponds to the storage memory in P-ACO, and the number of ants to be replaced in the memory is determined by the replacement rate $r$ $(0 < r < 1)$. The ants to be replaced in step $t$ are selected in order of the total environmental path length. In more detail, the following three algorithms are used:

- Random immigrants ACO (RIACO)
  The paths of the migrating ants are randomly generated. When ants with random paths enter the memory, pheromones that were concentrated between cities near the optimal solution are dispersed, increasing the diversity and allowing a wider range of search. Therefore, RIACO is effective when the environment changes frequently and significantly. On the other hand, randomly generated paths make it difficult to centralize the search

using past information, which may hinder the search when the environment is static or when the changes are mild.

- Elitism-based immigrants ACO (EIACO)
  Migrating ants are generated based on past information. Specifically, when generating migrating ants at time step $t$, the best ants in $P(t-1)$ are used as past information. The inver-over operator was used for a local search method just as used in GA, and it can be used to generate a new solution (i.e., Hamilton path) by replacing a part of the permutation in the reverse order. It also uses information from other individuals in the population (see [138] for details). Unlike RIACO, EIACO is useful when environmental changes are infrequent or mild. Elitism-based immigrants can help to extend their search a bit, making it easier for them to adapt to new environments. However, when changes in the environment are large, the best ant in $P(t-1)$ does not always perform well in the new environment, so the search range is not extended enough, and it is difficult to avoid the risk of falling into a local optimum.

- Memory-based immigrants ACO (MIACO)
  In addition to short-term memory, long-term memory $K_{\text{long}}(t)$ (size $k_l$) is used to generate migrating ants. This has the role of retaining information about the past environment. Unlike $K_{\text{short}}(t)$, $K_{\text{long}}(t)$ is updated whenever a change in the environment occurs. The process is also different when the memory is already full, replacing the ants in $K_{\text{long}}(t)$ that are closest to the optimal ants before the environment change. The proximity of ants is defined as the number of common edges among the solutions of ants. The ants in $K_{\text{long}}(t)$ are responsible for monitoring the environment, re-evaluating it at every step and acting as change detectors. EIACO is effective in dynamic TSPs where environments similar to past environments reappear, i.e., where similar environments appear periodically, because EIACO can retain information about the changing environment over time.

## 3.7 Creating new meta-heuristics by means of LLM

In this section, we will have LLMs[8] improve existing metaheuristics or propose new ones. Then, let LLMs output those codes as well. We run those codes and see how effective they are. The following description and experiments are based on [99].

The generation of new metaheuristics follows steps described below:

**Step 1** Ask LLMs to provide three to six examples of meta-heuristics and explain them in detail.

[8]ChatGPT-3.5 was used in the following experiments.

**Step 2** For each of the algorithms listed in **Step 1**, have the LLMs explain the important features and factors that affect search performance. Discuss each method's validity. Pay attention to "hallucination".

**Step 3** For a specific task, have the LLMs propose a hybrid version of the metaheuristics listed in **Step 1** or an improved version of them. Then, have the LLMs explain them and output pseudo-code or `python` code. As prompts, it is preferable to include important elements for search, such as "Maintain diversity," "Strike an effective balance between exploration and exploitation," "Do not fall into local solutions," and "Minimize the number of times the objective function is evaluated."

**Step 4** Verify the LLM proposal by actually running the pseudo-code, observing whether there are fatal errors (clear bugs) and whether they can be corrected. In addition, discuss the performance at the execution level, computational efficiency, in comparison with existing metaheuristics, etc.

Specific prompts and LLMs' responses are explained below.

> **Prompt 1**: "Provide a list of six swarm-based metaheuristics that have shown promising performance in global optimization tasks, and their unique characteristics make them ideal candidates for generating hybrid optimization algorithms using large language models."

> **Prompt 2**: "Provide reasoning for the selection of these algorithms."

> **Prompt 3**: "Identify unique components of each algorithm. Provide a description of each component and reason how it affects the algorithm performance and population dynamic."

These are the prompts for the above-mentioned **Step 1** and **Step 2**. In response to **Prompt 1**, LLM responded with the following metaheuristics:

- Particle Swarm Optimization (PSO)
- Ant Colony Optimization (ACO)
- Firefly Algorithm (FA)
- Bat Algorithm (BA)
- Cuckoo Search (CS)
- Bees Algorithm (BA)

For **Prompt 2** and **Prompt 3**, LLM accurately described the remarkable features of each algorithm, but included some incorrect responses, such as:

- PSO's concept of particles moving through the search space can be easily combined with the capabilities of large-scale language models to generate a variety of candidate solutions and to incorporate domain-specific knowledge.
- BA utilizes textual information to guide the search process and can be combined with a large language model to generate a variety of candidate solutions.

As can be seen from this, hallucination is occurring by interpreting the task of devising metaheuristics as integrating meta-heuristic algorithms and large-scale language models. Overall, the wrong responses were pulled by the phrase "using large language models" in the first prompt.

We then asked LLM to respond with the following prompt to generate improved metaheuristics.

**Prompt 4**: "Suggest a metaheuristic algorithm that would incorporate some of these components with the aim to promote population diversity and balance exploration and exploitation effectively."

**Prompt 5**: "Provide a description of the proposed ESEEO (Enhanced Swarm Exploration and Exploitation Optimizer) algorithm."

**Prompt 6**: "Suggest a metaheuristic algorithm that would incorporate some of these components and was effective for expensive optimization tasks, where the number of objective function evaluations is limited."

**Prompt 7**: "Provide a description of the proposed LESO (Limited Evaluation Swarm Optimizer) algorithm."

**Prompt 8**: "Provide the pseudo-code of ESEEO algorithm."

**Prompt 9**: "Provide working python implementation of ESEEO algorithm."

**Prompt 10**: "Provide the pseudo-code of LESO algorithm."

**Prompt 11**: "Provide working python implementation of LESO algorithm."

**Prompt 12**: "Provide a reasoning for the design of ESEEO."

**Prompt 13**: "Provide a detailed step-by-step reasoning for the design of ESEEO."

**Prompt 14**: "Provide a reasoning for the design of LESO."

**Prompt 15**: "Provide a detailed step-by-step reasoning for the design of LESO."

The actual codes are generated by **Prompt 9** and **Prompt 11**. ESEEO and LESO are considered to be obtained under the same conditions as in [99]. See Appendices B.1 and B.2 for details of the codes obtained. For the sake of comparison, we tried to solve the problem of minimizing the benchmark function using the existing method PSO and the LLM devised methods, i.e., ESEEO and LESO. The benchmark functions used are Ackley, Griewank, Schwefel, and Xin-she-yang functions from [172]. The shapes and definitions of these functions are shown in Fig. 3.40, Table 3.8 and Table 3.9. As for the minimization, we set the search time of each method to be about the same.

The experimental results are shown in Table 3.10. The mean, the variance, and the best obtained values over 30 runs are shown in the table. It can be seen that PSO gives mostly optimal values, while ESEEO is ineffective due to the large variation in solutions. Although LESO arrives at a relatively good solution, it falls into a local valley.

Considering the above results, we again asked LLMs to propose a new meta-heuristics. This time, the following points were improved.

**Table 3.8:** Benchmark functions from [172].

| **Function name** | **Definition** | **Ranges** |
|---|---|---|
| **Ackley** | $-20\exp(-0.2\sqrt{\frac{1}{30}\sum_{i=1}^{30}x_i^2}) - \exp(\frac{1}{30}\sum_{i=1}^{30}\cos 2\pi x_i)$ | $[-32,32]^{30}$ |
| **Griewank** | $\frac{1}{4000}\sum_{i=1}^{30}x_i^2 - \prod_{i=1}^{30}\cos(\frac{x_i}{\sqrt{i}}) + 1$ | $[-600,600]^{30}$ |
| **Schwefel's P.2.26** | $-\sum_{i=1}^{30}(x_i \sin(\sqrt{\mid x_i \mid}))$ | $[-500,500]^{30}$ |
| **Xin-she-yang** | $(\sum_{i=1}^{n}\mid x_i \mid)\exp\left[-\sum_{i=1}^{n}\sin(x_i^2)\right] \quad (n = 2 \sim 30)$ | $[-2\pi,2\pi]^n$ |

**Table 3.9:** Minimum values for benchmark functions from [172].

| Function name | Optimum |
|---|---|
| **Ackley** | $F(0,\cdots,0)=0$ |
| **Griewank** | $F(0,\cdots,0)=0$ |
| **Schwefel's P.2.26** | $F(420.9678,\cdots,420.9678)=-12569.5$ |
| **Xin-she-yang** | $F(0,\cdots,0)=0$ |

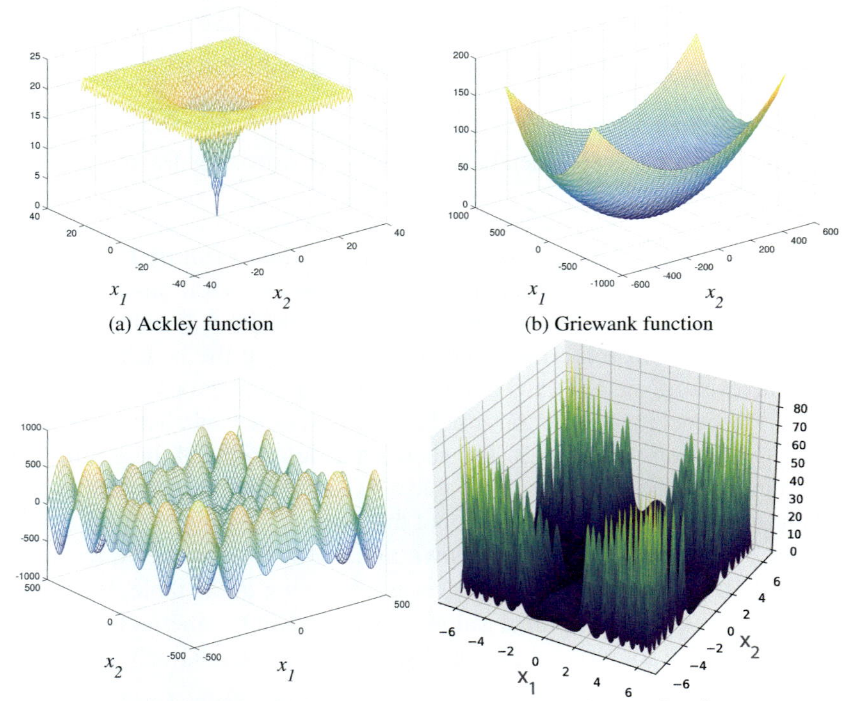


(a) Ackley function

(b) Griewank function

(c) Schwefel's Problem 2.26

(d) Xin-she-yang function

**Figure 3.40:** Benchmark functions from [172].

1. Remove the phrase "using large language models" from **Prompt 1**. This is because the above results suggest that this phrase is noise.

2. Eliminate similar metaheuristics methods among the proposed methods.

3. Add the prompt "maintain diversity, strike a balance between exploration exploitation and exploitation exploitation, and avoid falling into locally optimal solutions."

**Table 3.10:** Comparison of search performance of different metaheuristics.

| | | PSO | leso | eseeo | DEHO | merge |
|---|---|---|---|---|---|---|
| **Ackley** | Ave. | **1.56E-06** | 13.4 | 11781.7 | 1.43E-02 | 2.94E-03 |
| | Std. | 1.53E-06 | 36.3 | 14243.4 | 7.59E-03 | 5.70E-03 |
| | Min. | 1.98E-08 | -1.1 | 1078.2 | 2.30E-03 | **0.0** |
| **Griewank** | Ave. | 4.43E-03 | 5091.8 | 38643.1 | 6.67E-03 | **3.44E-03** |
| | Std. | 3.62E-03 | 5027.4 | 66271 | 7.31-E03 | 3.69E-03 |
| | Min. | 1.66E-15 | 0.4 | 551.1 | 1.62E-07 | **0.0** |
| **Schwefel** | Ave. | -7.8 | 6556.3 | 8183.1 | -172.5 | **-1951.4** |
| | Std. | 2.66E-16 | 5325 | 10842.2 | 256.3 | 3900.9 |
| | Min. | -7.8 | 182.1 | 7.8 | -1153.7 | **-20853.5** |
| **Xin-she-yang** | Ave. | **5.41E-06** | 515.4 | 19612.2 | 6.88E-03 | 2.14E-03 |
| | Std. | 1.67E-05 | 1499.7 | 27908.2 | 3.33E-03 | 2.56E-03 |
| | Min. | 7.09E-08 | -1.5 | 61.4 | 9.54E-04 | **1.17E-55** |

After making improvements in points 1 and 2 above, the proposed metaheuristics that maintained the most diversity were Bee Colony Optimization (BCO), Particle Swarm Optimization (PSO), Ant Colony Optimization (ACO), Cuckoo Search (CS), Genetic Algorithm (GA), and Differential Evolution (DE). For the third point, we prompted the following statements when proposing the metaheuristics and had LLM devise the algorithm:

> "Suggest a metaheuristic algorithm that would incorporate some of these components with the aim to promote population diversity and balance exploration and exploitation effectively. Please pay attention to the following points: 1. Maintain diversity of solutions. 2. Achieve an effective balance between exploration and exploitation. 3. Develop an algorithm that does not fall into local optima."

In this way, LLMs generated an algorithm called Diversity-Enhanced Hybrid Optimization (DEHO). The generated DEHO code is shown in Appendix B.3. DEHO combines the following methods in the `run` method:

- Bee Algorithm's waggle dance
- Cuckoo Search's extensive solution search with Levy flights
- velocity update of PSO, and
- fine-tuning with the hill-climbing method in `optional_local_search`.

DEHO's search performance is shown in Table 3.10. DEHO is able to generate a wide variety of candidate solutions. However, PSO produced better values with a smaller range of solution variation. Experimental results have shown that

DEHO is good at searching a wide area roug hly, whereas PSO is good at fine-tuning the solution. Effective integration of PSO and DEHO is expected to enable large scale search. If the range of initial solution candidates (initial population) is broadened by DEHO, it is possible to search for a large-scale solution. Therefore, we asked LLMs to propose a method called "merge" heuristics, in which DEHO is used to find the range of candidate solutions and PSO is used to fine-tune the range. The objective is to take advantage of both characteristics. The source code proposed by LLMs is shown in Appendix B.4.

The following is a concrete implementation code written for this purpose. The merge function can be executed without error with only a few lines of addition and modification.

```
deho = DEHO(objective_f, max_iterations, population_size,
            search_space_dimension, search_space_boundaries)
best_solution, best_fitness = deho.run()
pso = PSO(objective_f, dimension, num_particles,
                 np.min(best_solution), np.max(best_solution),
                 max_iterations, inertia_weight,
                 cognitive_weight, social_weight)
# DEHO determines the search range of candidate solutions
pso.global_best_fitness = best_fitness # initial best fitness
pso.global_best_position = best_solution # init. best solution
best_solution, best_fitness = pso.optimize()
```

The experimental results are shown in Table 3.10. The merge method is superior for all PSO and DEHO functions. Moreover, except for Xin-she-yang function, the solutions obtained are not far from the optimal values, which indicates the stability of the merge search.

As the next experiment, we asked LLMs to answer the questions from the first prompt, assuming that they have to solve the knapsack problem described in Section 2.5. However, since the knapsack problem is well-known, we adjusted the wording of the question to avoid memorized answers. As a result, each time LLM was used, the responses were different and different methods were generated.

For example, in one case, a combination of PSO and SA was generated, which would have been difficult for a novice to come up with and implement. The main parts of the final generated code are shown below:

```
def solve(self):
    current_temperature = self.initial_temperature

    for iteration in range(self.max_iter):
        for i in range(self.num_particles):
            # Update of PSO particles
            self.update_velocity()
            self.update_position()
```

```
                # Local search by SA
                current_solution = self.simulated_annealing(
                    current_temperature, self.particles_position)

                # Fitness evalution
                current_distance = self.calculate_total_distance(
                    current_solution)

                # Update on global best
                if current_distance < self.global_best_distance:
                    self.global_best_solution = np.copy(
                        current_solution)
                    self.global_best_distance = current_distance

            # Temperature update
            current_temperature *= self.cooling_rate

        return self.global_best_solution, self.global_best_distance
```

When run on the traveling sales problem (see Section 3.6), LLMs proposed a method combining SA and PSO as well. There were no errors or bugs in the execution, and the execution time was very short, so the computational efficiency was good. Since both PSO and SA have high processing speed, it is thought that hybridization can add diversity with SA while maintaining a certain level of accuracy with PSO without reducing the speed.

Furthermore, for the *N*–queen problem (Section 2.5), LLMs proposed an unexpected hybrid version that combines ACO and ABC. From the implementation of the algorithm to the solving part of the problem, we were able to run it without a single modification. On the other hand, LLMs would not have incorporated the output of the solution if it had not been explicitly stated in the prompt at the beginning. This confirmed the importance of how the prompt was written. From these results, we can say that it is useful enough that LLMs generate a working code for now. One drawback is that the performance may not be as expected if the parameters are used in their initial state, since the operation has not been checked by LLMs.

The above experiments confirm that it is possible for LLMs and people to collaborate in devising good algorithms. If LLMs are generating hybrids by taking into account the merits of each method, as in the integration of SA and PSO, this is an excellent achievement in terms of creating metaheuristics. On the other hand, given that there is little mention of less well-known metaheuristics, it seems that the LLM is still weak when it comes to utilizing minor or completely novel ones and making them more versatile.

The drawback, however, is that LLMs can propose "hallucinatory" meta-heuristics and that they do not actually understand them. Therefore, it may not be able to output the correct code, and we have to fix it manually. In such cases, human users will need to suggest specific steps.

In summary, with careful use, LLMs can be a useful tool in the creation of metaheuristics. Also, it seems to be easier to combine multiple ideas (on the form and text). There may be a possibility to come up with meta-heuristics ideas that are outside of the previous ideas.

# Chapter 4

# Emergence in Multiple Objectives

Computers are like humans - they do everything except think. (John von Neumann)

## 4.1 St. Petersburg Paradox

Consider the following game (bet):

- Keep tossing an unbiased coin until you get a face.
- You will receive \$2 for the first turn, \$4 for the second turn, \$8 for the third turn, ..., and $\$2^n$ for the $n$-th turn.
- How much should you pay to participate in this bet?

First, let us calculate the expected value of the money you will get from this bet. Of course, if the participation fee is less than the expected value, you should be willing to participate.

The expected value is calculated in the following way. The probability of getting a head for the first $n$ times is the probability of getting tails up to $n-1$ times, and the probability of getting a head the next time. Since it is a fair coin, the probability of getting a head or a tail on each try should be $1/2$. Therefore, the expected value is \$$2^n$ for the $n$-th try, so the participation fee becomes infinite as follows:

$$
\begin{aligned}
\text{Average amount won when tossing a coin } n \text{ times} &= \sum_{k=1}^{n} 2^k \times \frac{1}{2^k} = \sum_{k=1}^{n} 1 = n \\
\text{The expected value of bet} &= \lim_{n \to \infty} n = \infty
\end{aligned}
$$

But is this true? No rational person would pay unlimited money to take this bet. Something is wrong.

Daniel Bernoulli[1] answered this question. His insight was to see that the utility of money is not the amount of money itself. He assumed that the utility of money is like the logarithm of the amount of money in current mathematics. The meaning of utility being a logarithmic function will be explained in detail later. Simply put, it means that the satisfaction (i.e., utility) derived from an increase in wealth is inversely proportional to the quantity of the good previously possessed. With this assumption, the expected value of a bet can be calculated as follows:

$$
\begin{aligned}
\text{Average utility when tossing a coin } n \text{ times} &= \sum_{k=1}^{n} \log(2^k) \times \frac{1}{2^k} \\
&= \log(2) \times \sum_{k=1}^{n} \frac{k}{2^k} \\
&= 2 \cdot \log(2) \times (1 - \frac{1}{2^n} - n \times \frac{1}{2^{n+1}})
\end{aligned}
$$

$$
\begin{aligned}
\text{Expected utility value} &= \lim_{n \to \infty} \{2 \cdot \log(2) \times (1 - \frac{1}{2^n} - n \times \frac{1}{2^{n+1}})\} \\
&= 2 \cdot \log(2) = \log(4)
\end{aligned}
$$

Thus, the above calculations show that this bet is only worth \$4. In other words, it is reasonable to conclude that it is OK to pay \$4 to participate, but not to participate if it is more than \$4.

Now let us consider what utility is, which is the theme of this chapter. It is defined as follows:

**Definition 4.1 Utility and preference**

- Utility is what an individual wishes to maximize when choosing one of many actions.
- Preference is the outcome of a choice.

---

1 
Daniel Bernoulli (1700–1782): Swiss mathematician and physicist. He is famous for Bernoulli's law in fluid mechanics. The second of three brothers, his elder brother Nicholas II and younger brother Johann II were also mathematicians and physicists. Known as the St. Petersburg Paradox because he lived in St. Petersburg, Russia.

Although somewhat archaic, this is the definition from utility theory by John von Neumann[2] and Oskar Morgenstein. They are famous for their work in game theory. Utility theory is also important research in artificial intelligence and cognitive science.

They define utility as a measure of the choice between relative relations. In other words, only the order is important. Since this is too abstract to handle, we consider the utility function as a quantitative index of the preference relation.

The utility function is defined as follows:

**Definition 4.2 Utility function**
The preference of a state $S$ is given by a real-valued function $U(S)$. In other words,

- If $U(S_i) > U(S_j)$, then $S_i$ is preferred to $S_j$.
- If $U(S_i) = U(S_j)$, then there is no difference between the two states $S_i$ and $S_j$ (i.e., no discrimination).

In this case, $U(S)$ is called "utility function."

Consider rational human behavior. What will a human do when the possible actions are $A, B, C, \cdots$? Usually, they choose the action that maximizes their expected utility ($EU$). The expected utility of the action $A$ is as follows:

$$EU(A) = \sum_i P(Res_i(A)) \cdot U(Res_i(A)), \tag{4.1}$$

where $Res_i(A)$ represents all possible states after performing the action $A$. $P(Res_i(A))$ is the probability of reaching such a state. In the same way, we can calculate $EU(B), EU(C), \cdots$ and choose the largest action among them.

When can we use the utility function? This has been studied in mathematical detail by von Neumann and Morgenstein.

Now let the preference be described as follows:

- $A \succ B$: $A$ is preferred over $B$. By the same token, $A$ is dominant over $B$.
- $A \sim B$: $A$ and $B$ are indiscriminate.
- $A \succeq B$: $A$ is preferred to $B$ or $A$ and $B$ are indiscriminate.

2 
John von Neumann (1903–1957): Hungarian-born mathematician in the United States. He conducted pioneering researches in a wide range of fields from physics to psychology and mathematics. Known as the "Father of Computers" for his advocacy of the Neumann-type computer. His posthumous book on a self-reproduction model using cellular automata was published after his death.

Suppose that the following six axioms are satisfied:

**Definition 4.3 Axiom of rational agents**

**Completeness** $(A \succ B) \vee (B \succ A) \vee (A \sim B)$

**Transitivity** $(A \succ B) \wedge (B \succ C) \implies (A \succ C)$

**Continuity** $(A \succ B \succ C) \implies \exists p[p,A;1-p,C] \sim B$

**Substitutability** $(A \sim B) \implies [p,A;1-p,C] \sim [p,B;1-p,C]$

**Monotonicity** $(A \succ B) \implies (p \geq q \iff [p,A;1-p,B] \succeq [q,A;1-q,B])$

**Decomposability** $[p,A;1-p,[q,B;1-q,C]] \sim [p,A;(1-p)q,B;(1-p)(1-q),C]$

Note that choosing $X$ with probability $p$ and $Y$ with probability $1-p$ is written as $[p,X;1-p,Y]$. To choose $X,Y,Z$ with probability $p,q,r$ is written as $[p,X;q,Y;r,Z]$ $(p+q+r=1)$.

When this axiom is satisfied, it can be shown that there exists a real-valued utility function $U$ described below:

$$U(A) > U(B) \iff A \succ B$$
$$U(A) = U(B) \iff A \sim B$$

Let us elaborate more on the rational agent axiom. Completeness is that for every state, it can be determined whether it is a dominance relation or a non-discriminatory one. Transitivity shows that if $A$ v.s. $B$ and $B$ v.s. $C$ are non-discriminatory, then $A$ and $C$ must also be non-discriminatory. Continuity also implies that if $A$ is preferred to $B$, then there is some real number $\alpha \in [0,1]$ such that $\alpha A$ (choosing $A$ with probability $\alpha$) and $B$ are indiscriminate.

As we will see later, the rational agent axiom is not always satisfied in real situations. For example, in Section 4.2 we will consider an example where transitivity does not hold.

A typical utility function is the logarithmic function[3] assumed by Bernoulli. Figure 4.1 shows an example of this utility function $U(x) = k\log(x)$ $(k > 0)$. This is the same as the following psychophysical law (see Fig. 4.2):

***Theorem 4.1***
**Weber Fechner law**
*Psychological sensation is perceived in proportion to the logarithm of the increase, not to the intensity of the stimulus.*

In other words, the sensation is logarithmic, even if the intensity of the stimulus is 10, 100, or 1,000 times greater. For example, the pleasure of receiving $100 is

[3]Also called as von Neumann–Morgenstern utility index.

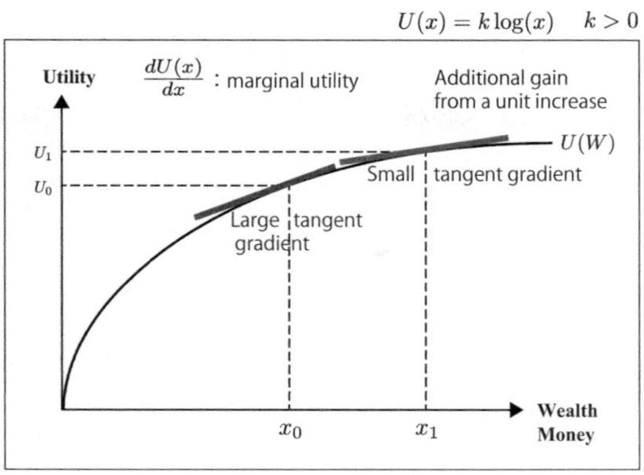


**Figure 4.1:** Example of a utility function.

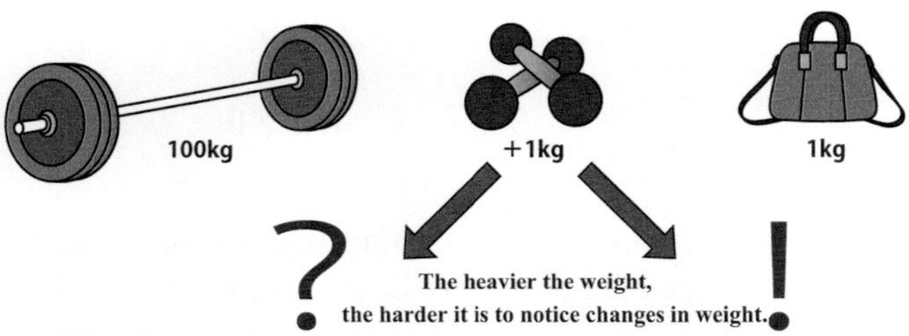


**Figure 4.2:** Weber–Fechner law.

different when you have \$100,000 than when you have \$10. Let us consider the slope $\frac{dU(x)}{dx}$ of the tangent line of the utility function $U(x)$ at $x$. This is the utility that increases with an increase of one unit (i.e., \$1), which is called "marginal utility."

As can be seen from Fig. 4.1, if $x_0 < x_1$, the gradient at $x_0$ is larger than at $x_1$. This means that the larger the amount of money, the smaller the marginal utility. In other words, when you receive the same amount of money, the smaller the amount of money you have, the greater the incremental utility and the greater your pleasure. This is true not only for the log function, but for any upward convex utility function. This is called "Law of diminishing marginal utility."

> ***Theorem 4.2***
> **Law of diminishing marginal utility**
> *If the utility function is upward convex, the larger the amount, the smaller the marginal utility.*

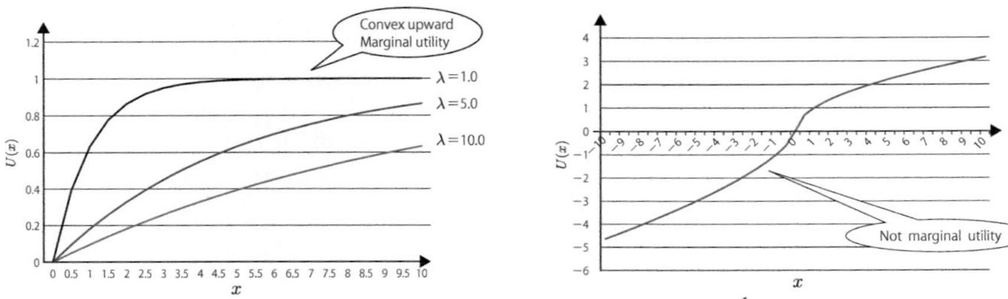


(a) $U(x) = 1 - e^{-x/\lambda}$ $(\lambda > 0)$ (b) $U(x) = \begin{cases} \sqrt{x} & \text{if } x \geq 0 \\ -\mid x \mid^{\frac{2}{3}} & \text{if } x < 0 \end{cases}$

**Figure 4.3:** Examples of utility functions.

In addition to the logarithmic function (see Fig. 4.1), other forms of utility functions are be used:

$$U(x) = 1 - e^{-x/\lambda} \ (\lambda > 0), \tag{4.2}$$

$$U(x) = \begin{cases} \sqrt{x} & \text{if } x \geq 0, \\ -\mid x \mid^{\frac{2}{3}} & \text{if } x < 0, \end{cases} \tag{4.3}$$

where eq. (4.2) is convex upward, i.e., diminishing marginal utility. On the other hand, eq. (4.3) is not marginal utility diminishing (see also Fig. 6.5).

We will explain how a person would cope with risk when the utility function is convex upward in a traditional model. Let $W$ in Fig. 4.4(a) be the current state (the total amount of assets currently held). Naturally, $U(W^*)$ is the value of the current utility. Now consider the following two types of bets:

**Bet1** 50-50 chance of gaining \$$h$ or losing \$$h$.
The expected utility of this bet is $U^h(W^*) = \frac{1}{2}U(W^* + h) + \frac{1}{2}U(W^* - h)$.

**Bet2** 50-50 chance of gaining \$$2h$ or losing \$$2h$.
The expected utility of this bet is $U^{2h}(W^*) = \frac{1}{2}U(W^* + 2h) + \frac{1}{2}U(W^* - 2h)$.

As can be seen from Fig. 4.4(b), when the utility function is convex upward, the following inequality holds:

$$U(W^*) > U^h(W^*) > U^{2h}(W^*). \tag{4.4}$$

In general, the larger $h$ is, the smaller the expected utility of the bet. Thus, with this type of utility function, we prefer the status quo to a betting situation, and we also prefer small bets to large bets. In other words, a rational person would avoid betting, which is why it is called risk aversion. Figure 4.4(c) shows the current state ($W^*$) and the utility value $U^h(W^*)$ of **Bet1** for the risk-averse (upward convex) utility function. Note that $W''$ has the same utility value as **Bet1**. We should

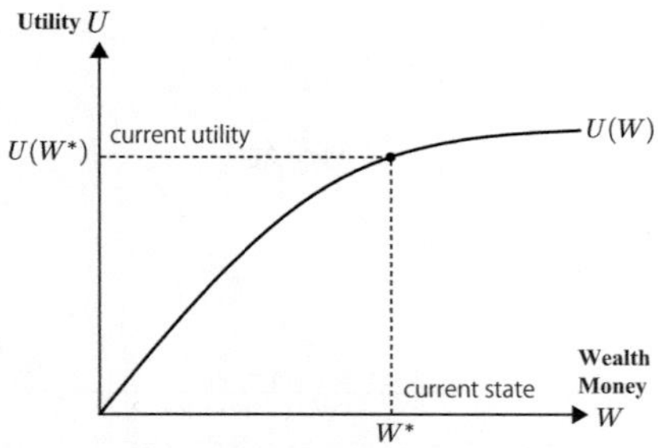


(a) Utility function is convex upward.

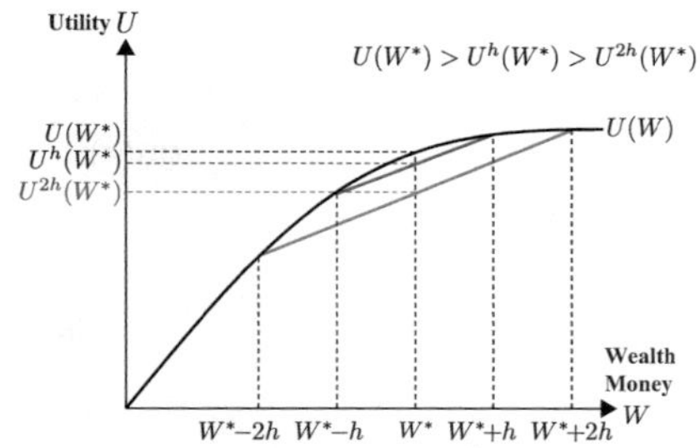


(b) Expected utilities of **Bet1** $U^h(W^*)$ and **Bet2** $U^{2h}(W^*)$.

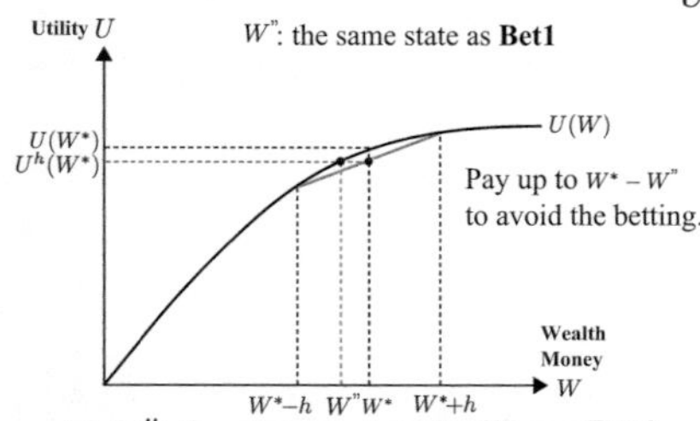


(c) $W''$: the same expected utility as **Bet1**.

**Figure 4.4:** Utility functions of risk aversion.

pay to avoid taking the bet if the amount is up to $W^* - W''$. This is why many people take out insurance, as will be explained in detail below.

Next, let us consider the case where the utility function is convex (see Fig. 4.5). In this case, the same calculation yields the following inequality:

$$U(W^*) < U^h(W^*) < U^{2h}(W^*). \tag{4.5}$$

In other words, the expected utility of the betting is higher than the status quo, and the larger the bet, the better. Therefore, it is a risk-seeking type.

In summary, rational human behavior can be divided into three types depending on whether one likes or dislikes risk (see Fig. 4.6). This is determined by whether the utility function is upward convex, downward convex, or linear. The three types are risk-seeking, risk-averse, and risk-neutral, respectively.

Why do people buy insurance? A risk-averse utility function will always avoid (fair) bets. As mentioned above, this is a case of diminishing marginal utility, and one may pay a certain amount to avoid the gamble.

Consider a person with a property worth \$100,000. He buys a \$20,000 car. Let us assume that there is a statistical probability of 25% that the car will be accidentally stolen or damaged. If this person follows von Neumann–Morgenstern utility index $U(W) = \log(W)$, then the expected utility can be given as follows:

$$\begin{aligned} E(U) &= 0.75 \times U(100,000) + 0.25 \times U(80,000), && (4.6)\\ &= 0.75 \times \log(100,000) + 0.25 \times \log(80,000), && (4.7)\\ &= 4.975. && (4.8) \end{aligned}$$

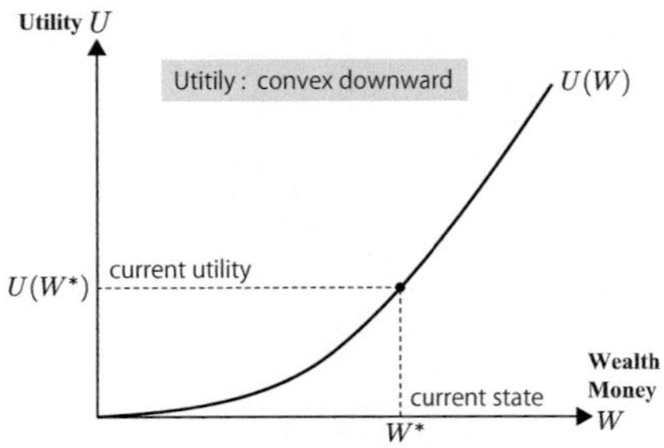


(a) Utility function is convex downward.

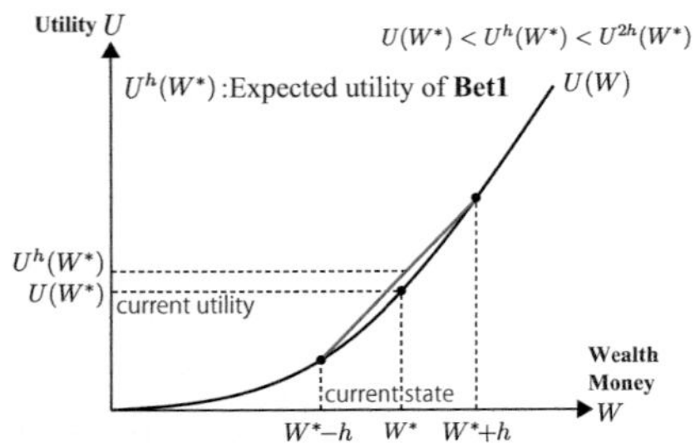


(b) Expected utilities of **Bet1** $U^h(W^*)$ and **Bet2** $U^{2h}(W^*)$.

**Figure 4.5:** Utility functions of risk seeking.

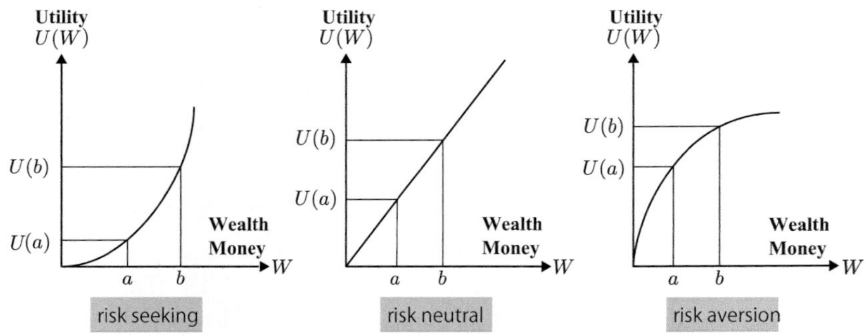


**Figure 4.6:** Do you like or dislike risk?

The first term indicates that the car has not been damaged by accident or theft, whereas the second term means the car damage by accident or theft, resulting in a loss of value. In this situation, the premium for a fair insurance policy would be \$5,000 (=25% of \$20,000). However, if you are risk averse, you should pay more than \$5,000 to avoid a gamble. Then, how much should you pay?

To find this, we need to subtract the utility value by the amount paid. If the premium is $x$, the utility is $E(U) = U(100{,}000 - x) = \log(100{,}000 - x)$. The maximum value of $x$ is equal to 4.975 according to eq. (4.8). Thus, we have the following equation:

$$\begin{aligned} 100{,}000 - x &= 10^{4.975}, \\ x &= 5593.9. \end{aligned}$$

This means that the maximum premium is approximately \$5594. Therefore, you should pay the premium if it is less than this value.

## 4.2 Pareto-efficiency and human intelligence

In the following discussion, the word "utility" is defined as the value representing the degree of preference of an individual. This is commonly regarded as the fitness of an objective function in evolutionary computation or swarm intelligence.

Let us look at the following comical anecdote from Sidney Morgenbesser, professor of philosophy at Columbia University (see Fig. 4.7):

> Sidney is told by a waitress after dinner that he can choose between an apple pie and a blueberry pie. He says "I want an apple pie." Shortly thereafter, the waitress comes back and says that a cherry pie is also an option. Sidney says "In that case I'll have a blueberry pie."

Why is this story funny? It is because the independence from irrelevant alternatives was lost. It might have been simply ignored if this was not said by a famous philosopher. However, one case actually occurred. During the 1995 World Championship of female figure skating, the preliminary ranking after the performance by three skaters was: 1st: Chen Lu (China), 2nd: Bobek (USA), and 3rd: Bonaly (France). Following this, Michelle Kwan (USA) performed and ranked in fourth place. As a result, the second and third places in the ranking were swapped and became: 1st: Chen Lu (China), 2nd: Bonaly (France), and 3rd: Bobek (USA). It was understood that there was no problem with the calculation of scores by the judges [100].

It is not only humans who lose the independence from irrelevant alternatives. There have been reports on an interesting behavior by a microorganism, called the slime mold [140]. A single cell of slime molds can move at the speed of a few centimeters per hour in the form of an amoeba. It can also transform into a

**Figure 4.7:** Independence from irrelevant alternatives.

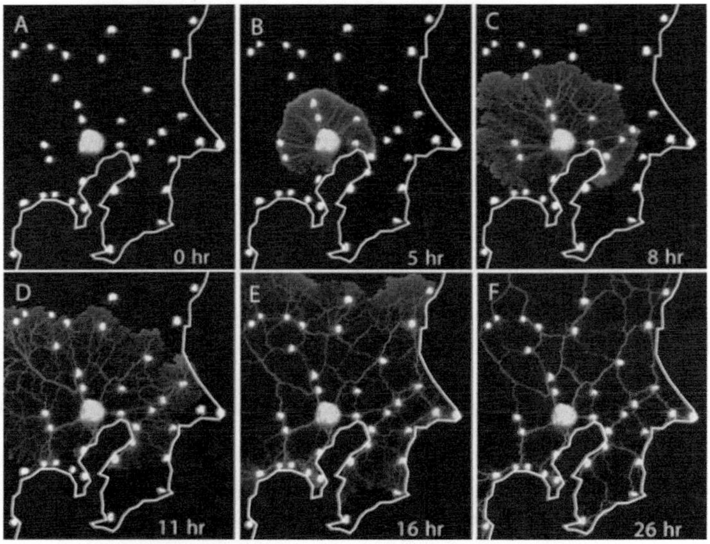


**Figure 4.8:** Transportation route by slime molds [140].

mushroom-shaped fruiting body (plasmodium) when thousands of cells congregate. Even though this plasmodium does not have a neurotransmitter system, it reacts to various external information, such as light and feeding stations (oats), and displays unified behavior. Consequently, it is able to go through a labyrinth in the shortest route or to form an optimal transportation route that appears as though it was designed by a human (Fig. 4.8, see also [55]). This is called "slime intelligence" and is a type of swarm intelligence, i.e., the subject of this book.

This slime mold was made to select how to be fed with oats [74]:

- $5g$ of oats in a bright spot,
- $3g$ of oats in a dark spot.

Slime molds dislike bright places. Hence, they chose between these two options at a 50/50 rate. Particularly, these two options were equally attractive to the slime mold. The fact that changing the amount of oats in the bright spot to 10g resulted in the slime mold choosing that option every time affirms this. However, when it was given the following three options, the results were odd (see Fig. 4.9):

- $5g$ of oats in a bright spot,
- $3g$ of oats in a dark spot,
- $1g$ of oats in a dark spot.

Even in this case, if considered rationally, the top two options should be selected at a fifty-fifty ratio. However, the slime mold changed its preference and

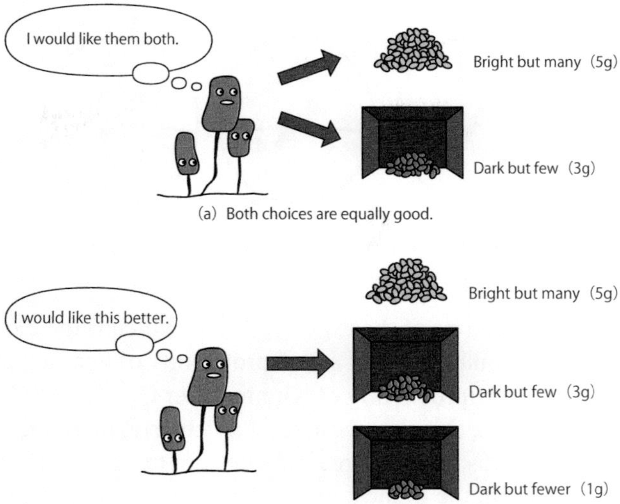


(a) Both choices are equally good.

(b) A new choice will change preferences.

**Figure 4.9:** Slime molds' preference.

selected the $3g$ in a dark spot at the rate of more than thrice that $5g$ in a bright spot. This result indicates the violation of independence from irrelevant alternatives.

Kenneth Arrow[4] demonstrated that the "majority rule," which is the fundamental principle of democracy, can be arbitrarily manipulated. This phenomenon, which shakes the foundation of democracy, can be considered to have emerged from the irrationality of a society composed of multiple agents. Particularly, even if each individual acts rationally, it does not necessary result in the maximum benefit to the society as a whole.

Let us introduce mathematical paradox, which can be explained more difficultly. There are three roulettes, as shown in Fig. 4.10. One can win the prize (integer between 1 and 6) placed at the random spot where the roulette needle comes to rest when their wheels are turned. The small actual values are the area size of that spot. Thus, with roulette B, the probability is as follows:

- 0.56 probability of winning two points,
- 0.22 probability of winning six points,
- 0.22 probability of winning four points.

Which roulette should you place your bet on?

---

4  Kenneth Arrow (1921–2017): American economist. One of the most important figures in the history of 20th century economics. At the age of 51, he was the youngest person ever to win the Nobel Prize in Economics in 1972.

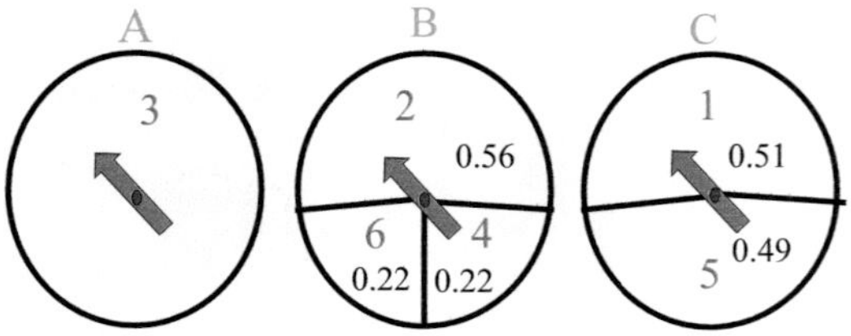


**Figure 4.10:** Which roulette should you place your bet on?

B is the best bet in terms of the expected value, which is the most understandable number. On the contrary, A has a 0.56 probability of winning over B (B loses if it gets two) and a 0.51 probability of winning over C. B has a 0.617 probability of winning over C; thus, C is the worst bet. Particularly, A would be the best bet. However, if the rule was that all three roulettes are turned at the same time, and the roulette resulting in the highest number wins, C ends up being the best bet. Thus, the relative merits can significantly change according to the perspective.

The above discussion demonstrates that the transitivity seen does not necessarily hold true. In fact, there are cases where even though $x \succ y$ and $y \succ z$, it became $z \succ x$.[5] A more mathematical explanation is as follows. Suppose that there are two evaluation items, and they become the utility functions $u_1$ and $u_2$ when they are considered individually. Let $x$, $y$, and $z$ exist in a two-dimensional space shown in Fig. 4.11. The following equation becomes true under this circumstance:

$$x \succ_{u_1} y, \quad y \succ_{u_1} z, \quad \text{and} \quad z \succ_{u_2} x, \tag{4.9}$$

where the subscripted number indicates evaluation by that utility function. The fact that the evaluator did not notice the two-dimensionality of the scale and unconsciously switched to a different utility function at each time resulted in intransitivity. This theory is called the "multi-dimensional utility theory."

In the multi-dimensional utility theory, the indifference curve is defined in a multi-dimensional space. The points on this curve are characteristically equivalent with each other, regardless of the position. For instance, the indifference curve shown in Fig. 4.12 is drawn in a two-dimensional space with two utility functions.

We often wish to pursue two or more mutually conflicting goals (Fig. 4.13). In finance, return and risk are generally conflicting objectives. In assets management and investment, return is the profit gained through management and risk is the uncertainty and amplitude of fluctuations; all involving present assets and the possibility of a future specific gain on those assets. Risk can be gauged from price volatility (the coefficient of variation). High volatility means large price fluctuations. Ordinary volatility is generally expressed in terms of standard deviation.

[5]Remember that $x \succ y$ means that $x$ is preferred to $y$.

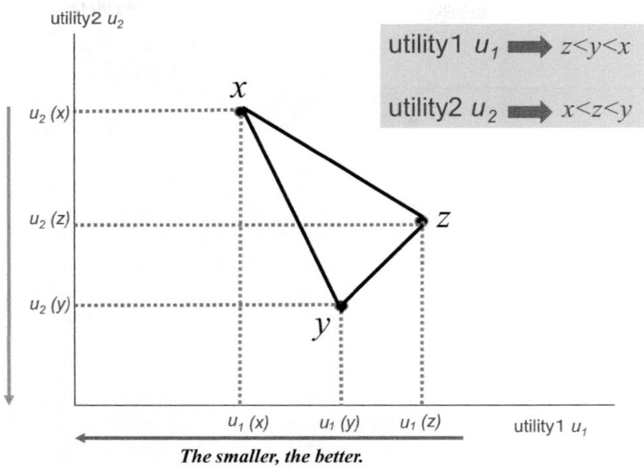


**Figure 4.11:** Multi-dimensional utility theory.

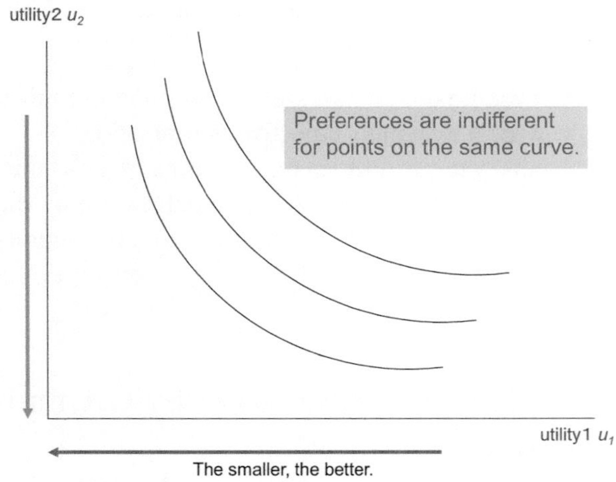


**Figure 4.12:** Indifference curves.

In general, risk and return are in a "trade-off" relationship. High-return, low-risk financial products would be ideal, but things are not actually that easy. Usually, the choice is one of the following two:

- Low risk, low return: The possibility of loss of the principal is low, but management returns will be small.
- High risk, high return: Losses may occur, but any returns might be large.

In considering the optimization problem, we have to this point limited the discussion to maximization of a single objective (the fitness function). As in risk

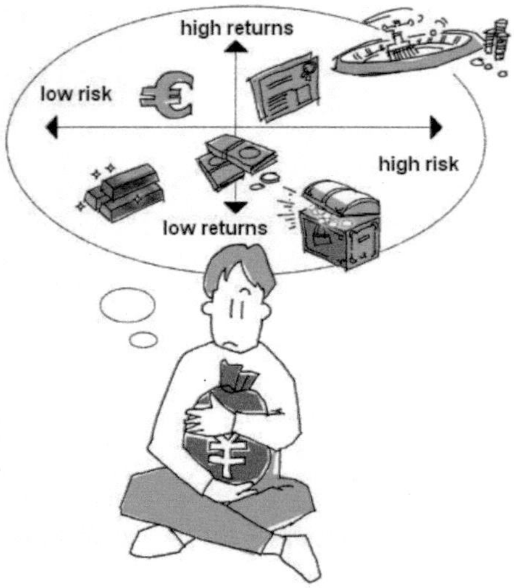


**Figure 4.13:** Risk vs. returns.

versus return, however, it is often necessary to consider optimization when there are multiple objectives, i.e., "multi-objective optimization."

In many problems, various utilities (e.g., returns vs. risk) must be simultaneously optimized. This is called multi-objective optimization. Many studies have tackled emergence from multi-objective optimization. In Section 4.3, we will see how to solve this multi-objective optimization by means of evolutionary computation and swam intelligence.

## 4.3 Multi-objective optimization by GA and swarm intelligence

In this section, we first introduce an approach based on evolutionary computation.

Before that, we will explain the motion generation of robots as an example of its application. For the motion generation of humanoid robots, the stability of motion is important, together with the accomplishment of the task that is its goal. Let us consider the design for the motion of kicking a ball. If the objective is to simply kick a ball as far as possible, an unstable motion that would result in the robot falling after the kick is enough. However, this will be problematic for the real robot usage. Thus, the trade-off between the ball distance and the motion stability should be considered. Figure 4.14 shows the result of designing the kicking motion by means of evolutionary computation. The plot in the lower figure shows the robot motion stability obtained through the evolution and the distance the ball flew. The stability is shown as the inverse of the maximum value of the

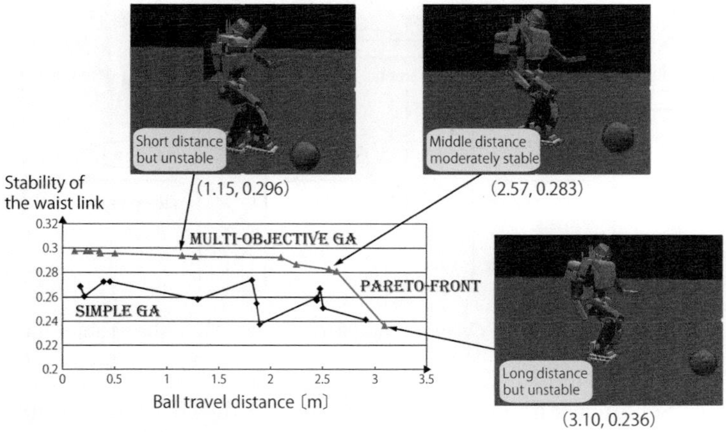


**Figure 4.14:** The motion generation of humanoid robots.

link speed of the hip. The larger these values are, the more desirable they become. Thus, this multi-objective optimization aims for the maximization of two value functions. In the multi-objective evolutionary computation, the solutions on the Pareto front[6] are simultaneously found. The figure shows an example of the representative motion of the solutions. The upper left solution represents a short movement distance of the ball and a stable motion, whereas the lower right solution represents a long movement distance and an unstable motion, where the robot is almost falling. The upper right solution represents moderate results in both movement distance and stability. In conventional optimization, one of them may be obtained by devising an objective function to be optimized. For instance, the lower polyline in the figure represents an example of the solution obtained through simple GA. Note that, in multi-objective optimization, different motions on the Pareto front are simultaneously discovered.

Pareto optimality are defined more formally as follows. Let two points $x = (x_1, \cdots, x_n)$ and $y = (y_1, \cdots, y_n)$ exist in an $n$-dimensional search space, with each dimension representing an objective (an evaluation) function, and with the objective being a minimization of each to the degree possible. The domination of $y$ by $x$ (written as $x <_p y$) are defined as

$$x <_p y \Longleftrightarrow (\forall i)(x_i \leq y_i) \wedge (\exists j)(x_j < y_j). \tag{4.10}$$

In the following, we will refer to $n$ (the number of different evaluation functions) as the "dimension number." A point that is not inferior to any other point will be called "non-dominated" or "non-inferior," and the curve (or curved surface) formed by the set of Pareto optimal solutions will be called the "Pareto front."

On this basis, it is possible to apply GA to multi-objective optimization. In the following example, we use the VEGA (Vector Evaluated Genetic Algorithm)

[6]Upper polyline in the figure. The detailed definition is given below.

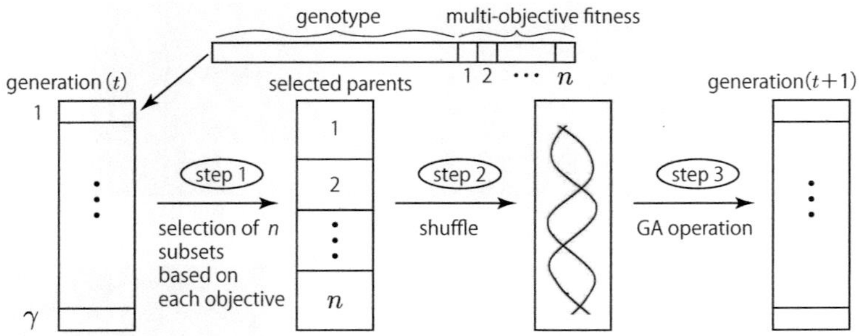


**Figure 4.15:** VEGA (Vector Evaluated Genetic Algorithms) algorithm.

system of Schaffer et al. [117, 116]. Selection with VEGA is performed as follows (see Fig. 4.15), where $n$ is the evaluation dimension number (the number of different evaluation functions):

**Step 1** Let the $n$ subsets be $Sub_i$ $(i = 1, \cdots, n)$.

**Step 2** In $Sub_i$, retain the individual selected only by evaluation function $i$.

**Step 3** Mix $Sub_1, Sub_2, \cdots, Sub_n$ and shuffle.

**Step 4** Produce the next-generation offspring, using the genetic operator on these sets.

Note that the selection is based upon every evaluation function (i.e., the function value in every dimension). In contrast, reproduction is performed not for each subset but rather for the complete set. In other words, crossover is performed for the individuals in $Sub_i$ and $Sub_j (i \neq j)$. In this way, we preserve the superior individuals in each dimension and at the same time select the individuals that are superior to the average in one or more dimensions.

Let us now consider the following multi-objective optimization:

$$\begin{aligned} F_{21}(t) &= t^2, \\ F_{22}(t) &= (t-2)^2, \end{aligned}$$

where $t$ is the only independent variable. The graphical appearance (i.e., "Pareto map") of this function is shown in Fig. 4.16. The objective is to select non-dominated points, such as those shown in the figure.

Figure 4.17 shows the results for generations 0 and 3. The population size was set to 30 individuals in each dimension. A crossover ratio of 0.95 and a mutation ratio of 0.01 were chosen. As shown in the figure, VEGA can effectively find a front (a Pareto front) containing no dominated points. However, it has lost several intermediate points.

VEGA selection, as may be seen in this example, involves a problem that may be described as follows. The selection pressure (in biological terms, the degree of

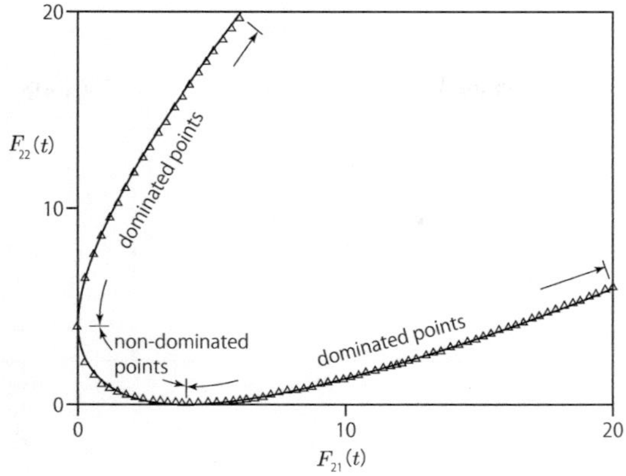


**Figure 4.16:** Pareto maps.

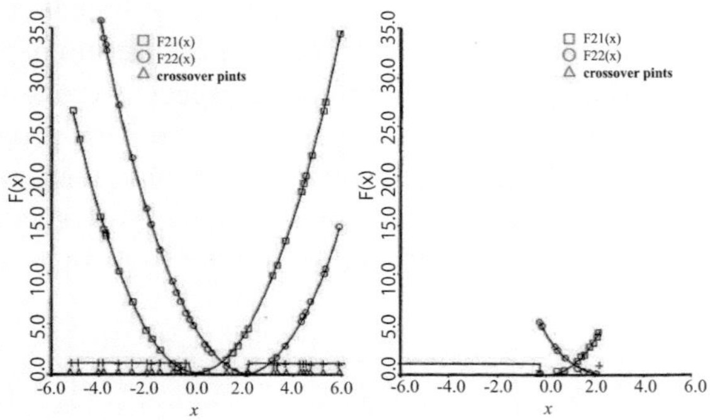


(a) Generation 0 (b) Generation 3

**Figure 4.17:** Results of multi-objective optimization [116].

selectivity applied to the survival and proliferation of the organism's population) on the optimum values in at least one dimension (evaluation function) works as desired. In the unlikely event that a utopian individual (an individual superior in all dimensions) exists, it may be possible to find it through genetic operation on the superior parents in only one dimension. In most cases, however, no utopian individual exists. It is then necessary to obtain the Pareto optimal points, some of which are intermediate in all dimensions.

In an ideal GA, it is desirable to have the same selection pressure applied on all of these. With VEGA selection, however, an intermediate point cannot survive. In effect, species differentiation occurs within the population, with dimension-specific superiority of each species. The danger of this is more prevalent when the Pareto optimal region is convex rather than concave (see Fig. 4.18). Two modifications have been proposed to resolve this difficulty. One is to apply an excess of heuristic selection pressure on the non-dominant individuals in each

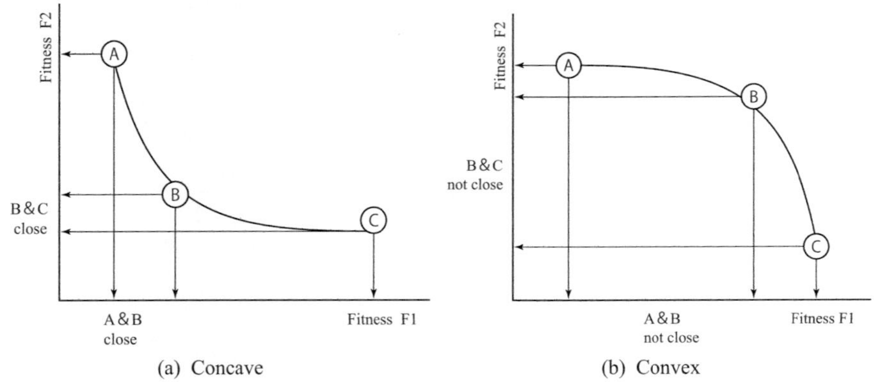


**Figure 4.18:** Pareto fronts.

generation. The other is to increase interspecies crossbreeding. Although ordinary GA selection is random, this is effective because utopian individuals are more readily produced by interspecies than intraspecies' crossover.

In addition to VEGA, many other methods have been proposed for Pareto optimization. For example, the ranking method assigns a rank $rank_i$ to each individual $i$ as an evaluation criterion. Let $n_i$ be the number of other individuals over which an individual $i$ is superior. Then the rank of $i$ is given as follows:

$$rank_i = 1 + n_i. \tag{4.11}$$

In VEGA, the evaluation criterion is a single function value, which has the disadvantage of bias in the solution. In particular, it is difficult to find a solution that has intermediate values in each objective function with VEGA. On the other hand, the ranking method has the potential to search for such solutions.

Another method is to consider niching. In this method, the degree of crowding around each individual is taken into account in the evaluation. Basically, while using the above ranking method, this method gives priority to the one with the smaller niching (lower congestion) for individuals of the same rank. Alternatively, this is achieved by penalizing individuals that are dense in the objective function value space. The penalty is based upon the number of individuals within a certain distance, and reduces the rank by that amount.

The distance $dist(x,y)$ between two individuals $x$ and $y$ is defined in the following way by using the maximum value $f_i^{max}$ and the minimum value $f_i^{min}$ of each objective function $f_1, \cdots, f_n$ in the population:

$$dist(x,y) = \sqrt{\left(\frac{f_1(x) - f_1(y)}{f_1^{max} - f_1^{min}}\right)^2 + \left(\frac{f_2(x) - f_2(y)}{f_2^{max} - f_2^{min}}\right)^2 + \cdots + \left(\frac{f_n(x) - f_n(y)}{f_n^{max} - f_n^{min}}\right)^2}, \tag{4.12}$$

where the values of each objective function $f_1, \cdots, f_n$ for $x$ and $y$ are $f_1(x), \cdots, f_n(x)$ and $f_1(y), \cdots, f_n(y)$.

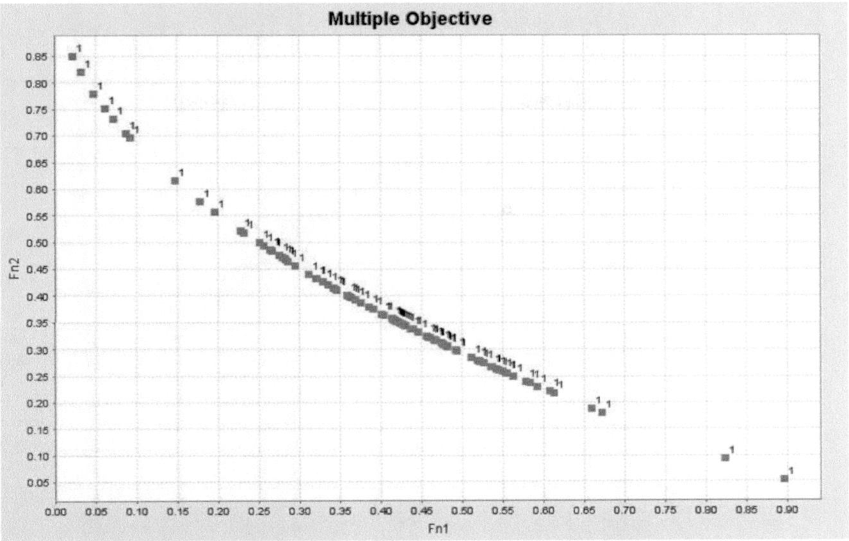


**Figure 4.19:** A convex Pareto front ($\alpha = 0.5$).

To test and experiment with evolutionary multi-objective optimization, a Pareto GA simulator is available (see Appendix A.2 for details).

As an example, let us try using the following functions as the fitness functions:

$$\begin{aligned} f_1(x) &= x, \\ f_2(x) &= 1 - x^{\alpha}. \end{aligned}$$

These functions will generate a concavity in the Pareto curve. In this problem, the Pareto front is convex when $\alpha = 0.5$ and non-convex when $\alpha = 2,0$. Figs. 4.19 and 4.20 show the results of execution with $\alpha = 0.5$ and $\alpha = 2,0$, respectively. The former clearly results in a convex Pareto front, and the latter in a concave Pareto front.

Let us now consider a more complex objective functions described below (see Fig. ??):

$$\begin{aligned} f_1(x) &= x, \\ f_2(x) &= 1 - x^{0.25} - x\sin(10\pi x). \end{aligned}$$

This is regarded as difficult problem, containing a discontinuous Pareto front. As the search progresses, the figures show that the individuals that were initially clumped together separate into a whole, forming a Pareto front consisting of diverse solutions.

Swarm intelligence is a diversity-oriented method, making it suitable for multi-objective optimization. Therefore, various metaheuristics have been developed for this purpose. For example, Figs. **??**(a) and (b) show the results of

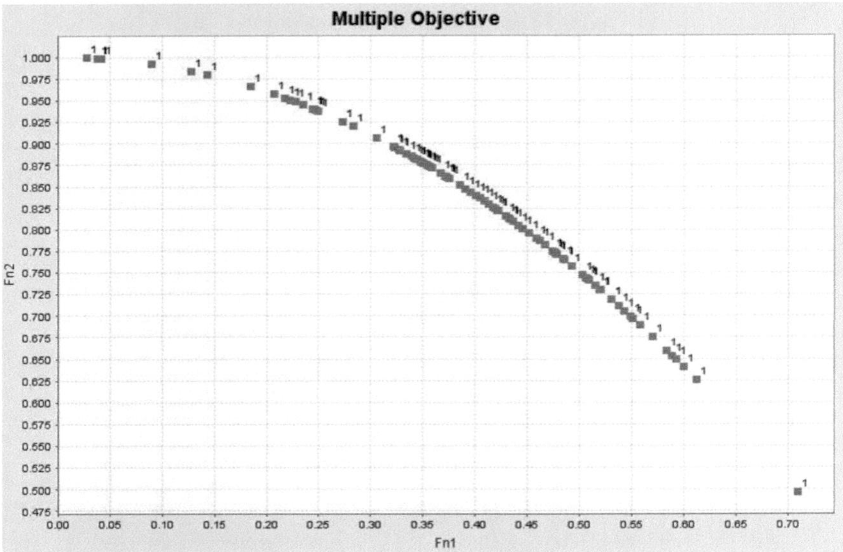


**Figure 4.20:** A concave Pareto front ($\alpha = 2.0$).

an ACO-based search. This is an example of the application of P-ACO method (see 101 page) to the functional optimization [92]. Figs. **??**(c) and (d) show the results of searching for discontinuous Pareto fronts using WOA (Whale Optimization Algorithm) and SLO (Sea Lion Optimization algorithm), respectively. Other methods such as PSO, Firefly, and cuckoo search are also employed for multi-objective optimization (see [57, Section 4.7] for details). However, there have been very few comparative studies of these diverse methods. Examining the robustness of various meta-heuristics for multi-objective optimization is an interesting topic for future research.

## 4.4 Criticism of indifference curve

Let us see the multi-dimensional utility function again. Ideally, the comprehensive evaluation is represented as the sum of each dimension. In the case of two dimensions, two items $X$ and $Y$ exist. If the utility functions are set as $u_1$ and $u_2$, the following equation should be true:

$$X \succ Y \iff u_1(X) + u_2(X) > u_1(Y) + u_2(Y).$$

The following conditions are required to satisfy this equation:

- Transitivity
- Comparability

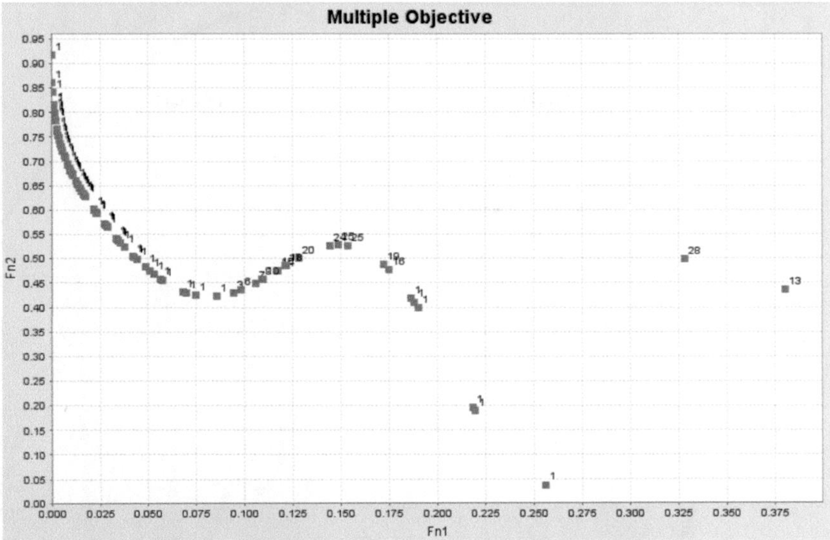


(a) Early stage of the search

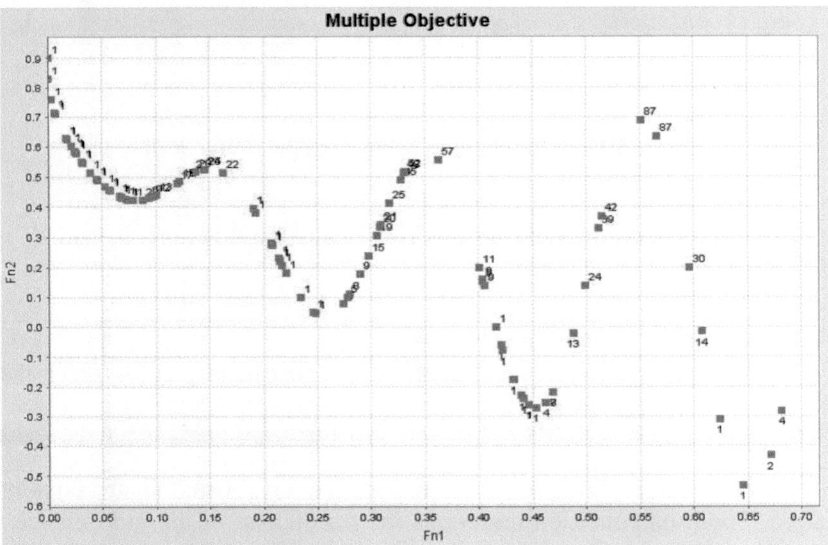


(b) End of search

**Figure 4.21:** Discountinuous Pareto fronts.

- Independence of preference
- Thomsen condition

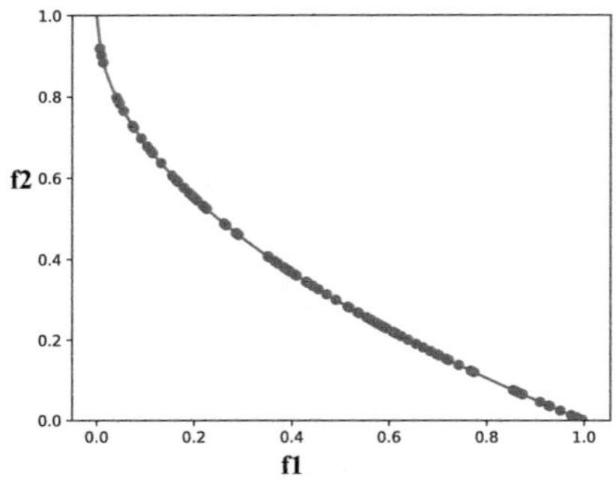


(a) Convex Pareto front by ants.

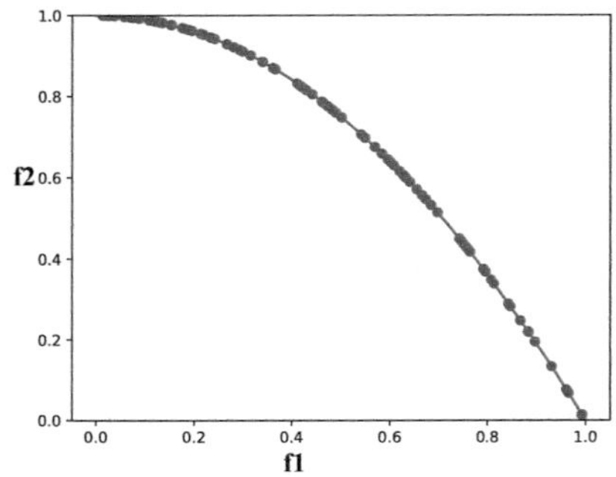


(b) Concave Pareto front by ants.

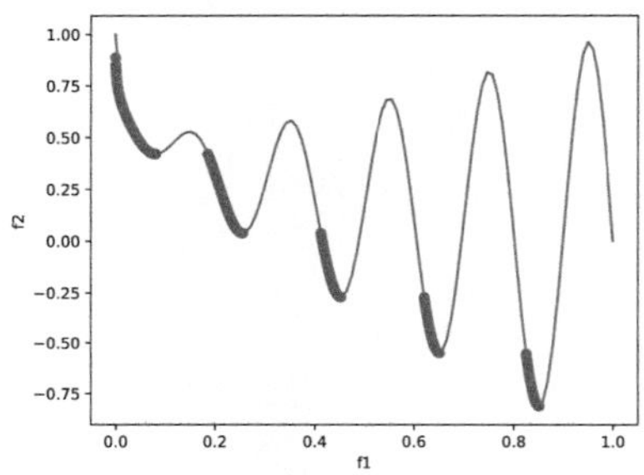


(c) Discountinuous Pareto front by whales.

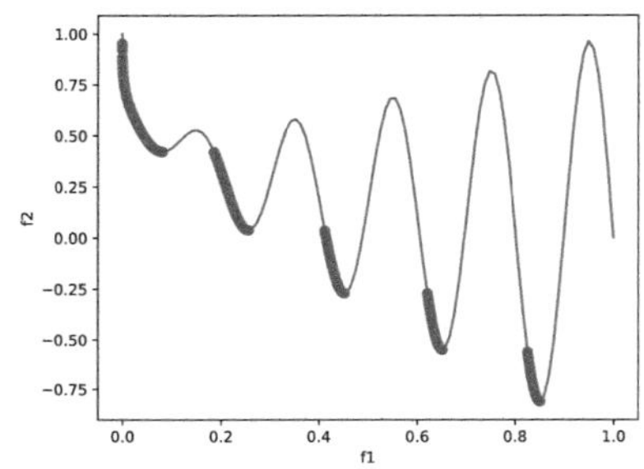


(d) Discountinuous Pareto front by sea lions.

**Figure 4.22:** Multi-objective optimization by swarm intelligence.

Thomsen condition indicates that two evaluation functions $X_1$ and $X_2$ satisfy the following:

**Definition 4.4 Thomsen condition**
For every $a,b,c \in X_1, p,q,r \in X_2$, if $(b,p) \sim (a,q)$ and $(c,p) \sim (a,r)$, then $(c,q) \sim (b,r)$ holds true.

Note that $A \sim B \Leftrightarrow A \succeq B$ and $B \succeq A$, i.e., $A$ and $B$ are equally preferred. Thomsen condition represents the parallelism of the indifference curve (see Fig. 4.23). Six candidate points can be found in the figure. In this case, Thomsen condition indicates that if two pairs with the same color are indifferent, the candidates of the remaining colors are also indifferent. For instance, if the candidates of black and white are both indifferent, the gray candidate would also be indifferent. If indifference curves can be drawn between the black points and between the gray points, an indifference curve can also be drawn between the two white candidates.

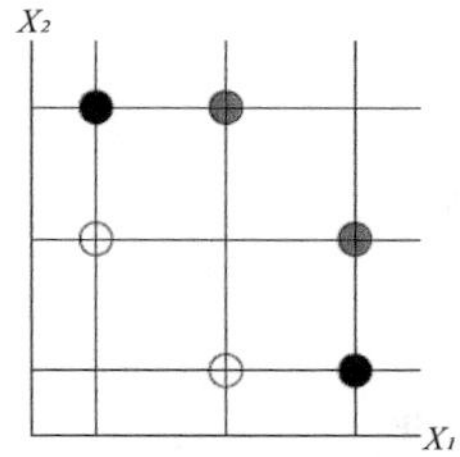


**Figure 4.23:** Thomsen condition.

This condition is generally unsatisfied. However, the sum of each dimension is sometimes used for convenience as the overall assessment value of each item.[7]

In contrast, the ways of thinking (e.g., indifference curve and Pareto optimization) have been criticized for their inability to explain human behavior well due to the fact they ignore the risk factor. The problem of the indifference curve is that it only represents the wealth and the money that is in possession at the end and does not indicate the current situation. Particularly, one cannot understand where humans are located from it. Demonstration experiments of selection without risk and decision making have proven that indifference curves are not given in the type of curve shown in Fig. 4.12.

Chapter 6 discusses solutions to these problems, particularly the realization of utilities that explain human cognition.

[7]The simple GA of the robot motion design (see Fig. 4.14) is also based on this condition.

# Chapter 5

# Emergence Applied to Deep Learning and Generative AI

Evolution is a stochastic hill-climbing algorithm, implemented in wetware. (*Uncertainty Isn't Always a Problem–It Can Be the Solution* by Ian Stewart)

## 5.1 Deep neural-evolution: overview

"Neuroevolution" is an active area of research in machine learning. This approach, also known as evolutionary artificial neural networks (EANNs) [171], integrates evolutionary methods with neural networks. The key feature of evolutionary neural networks is a genetic search for an optimal network. This saves the effort normally required to search for a neural network (such as constructing networks through trial and error).

Training via back-propagation, the method normally used to train neural networks, has been shown to often become stuck in only locally optimal solutions because it is based on the gradient descent method. To address this flaw, methods have been proposed that use evolutionary computation to learn which link weights to use. In other words, these methods express the link weights in a network as genotypes, and then use evolutionary computation to search for the best network. In this case, we derive the fitness value by the error in the output of the neural network (represented as a genotype, with lower errors representing better fitness). Methods for expressing link weights (genotypes) include binary strings and numeric vectors.

Note also that, when training a neural network, the network structure needs to be provided in advance. Neuroevolution, in contrast, can adaptively learn a network structure and size (number of nodes) appropriate to the task. Two different genotype representations have been proposed for driving the evolution of network structures.

**Direct coding** Directly represents the link state of the network structure

**Indirect coding** Encodes production rules for generating networks as genotypes

Indirect coding is closer to biological models than direct coding is. For example, direct coding might represent a network consisting of $N$ nodes $(n_1, n_2, \ldots n_N)$ as an $N \times N$ adjacency matrix, with each element taking a value of 0 or 1 such that a value of 1 for row $i$ and column $j$ indicates that there is a link from $n_i$ to $n_j$ and a value of 0 indicates that there is no such link.

For indirect coding methods, the focus of attention has been encoding for developmental systems. This approach uses observations from developmental biology as a basis to create abstract models of natural development in order to drive evolutionary computation algorithms. These models range from low-level cytochemistry through to high-level grammar-rewriting systems.

Once the genotype is determined, the following procedure is used to conduct training through evolutionary computations on network structures.

**Step 1** Find the phenotypes (the network structure of the EANNs) by decoding the genotypes for each individual.

**Step 2** Train each EANN. In this case, change the weights according to a predetermined training rule (such as back-propagation).

**Step 3** Derive the fitness value for each individual based on the training results. Normally, the squared error relative to the desired output is used for this.

**Step 4** Select candidate parents according to the fitness values, and then create the child generation by applying genetic operators to the selected parents.

**Step 5** Return to **Step 1** if the termination condition has not been met.

Furthermore, although Hebb's rule is the training rule typically used for neural networks, parameters such as the learning rate and momentum during back-propagation are not easy to tune because they are task dependent. Accordingly, there is active research into evolutionary methods for learning appropriate parameters as well as into using evolutionary computation to search for more general training rules than Hebb's rule.

A typical example of neuroevolution is NeuroEvolution of Augmenting Topologies (NEAT) [127], a method for efficiently optimizing the structure and parameters of neural networks. NEAT has demonstrated performance superior to that of conventional methods in a large number of problems. The NEAT method makes it possible to evolve variable network structures by gradually increasing the complexity of a low number of small networks with each successive generation.

Compositional Pattern-Producing Network (CPPN) is an ANN that takes coordinates as input and produces pixel values at corresponding points, generating regular geometric patterns. CPPN evolves to acquire more optimal and complex structures through the NeuroEvolution of Augmenting Topologies (NEAT) algorithm [128], grounded in GA. This process involves iteratively applying mutations (changes in edge weights or network structures) and crossovers, acquiring more optimal and complex neural network structures.

In several co-design methods for soft robots, CPPN is frequently used as the encoding scheme for genomes [20, 23, 24]. In these studies, CPPN's output is interpreted as the type of voxel at the given coordinates, generating the robot's morphology. Moreover, by representing controllers as CPPNs, the inheritance of controllers from parent to child becomes possible.

Tanaka et al. [136] proposed a co-design method that inherits both the structure and controller from parents to their children using the Hypercube-based NeuroEvolution of Augmenting Topologies (HyperNEAT) [129]. In HyperNEAT, CPPN receives two coordinates as input, and its output is interpreted as the weights between nodes corresponding to the two points, resulting in the generation of an ANN from CPPN. Tanaka et al., represented the robot's genome using a single CPPN. The CPPN generates two ANNs, where one is responsible for generating the robot's structure, and the other controls the generated structure. Based on fitness assigned depending on task achievement, the NEAT algorithm is applied to the CPPNs to create the next generation's CPPNs. This achieves co-design by inheriting both the structure and controller from parents to their children.

The emerging field of deep neuro-evolution is primarily concentrating on the design of optimal deep neural networks (DNN) automatically, which is called DNE (Deep NeuroEvolution). The manual design process is laborious and intractable at that scale of network design. Therefore, the current efforts are largely applied in DNN hyper-parameter optimization, architectural design, and weight training. Lately, some exciting applications of EC have been seen with other aspects of DL research.

The basics of DNE and its various applications are described in detail in [56].

In this chapter, we will discuss some applications of swarm intelligence and evolutionary computation to deep learning and generative AI. The researches discussed in the following sections are based largely on the literatures [71, 47, 143].

## 5.2 Image generation with diffusion model by IEC

Recent years have seen remarkable progress in image generation models. A technique for restoring images from noise, the denoising diffusion probability model, has been developed [52]. This was computationally intensive and required sophisticated facilities, but improvements have been made and the latent diffusion model [112] uses latent space for noise elimination, thereby significantly reducing the required computational resources.

As a result, anyone can now easily generate images using text-to-image generation models. However, it is difficult for users to come up with a prompt to generate the creative image because different people imagine different images for the same sentence.

In this section, we explore that the diffusion model can be made more usable for non-skilled users by exploring the latent space using IEC (see Section 2.2) instead of prompt tuning.

### *5.2.1 Stable diffusion model and IEC*

Recent advancements in image generation have seen the emergence of various techniques for restoring images from noise. Among these techniques are the denoising diffusion probability model (DDPM) [52] and the latent diffusion model (LDM) [112]. While the DDPM was computationally intensive and required sophisticated facilities, the LDM utilizes the latent space to eliminate noise, thereby significantly reducing the computational resources required.

Additionally, the stable diffusion model has been developed, which predicts images from text and is trained on billions of pairs of images and alt text on the Internet [112]. This model is based on a function that adds noise to the original image in the latent space, and the function reconstructs the image from the noise step by step to generate the image. The text is converted into a vector using a tokenizer, which is then converted into latent variables through an encoder (Fig. 5.1). First, the user inputs a prompt, which is randomly determined by the user. The prompt is converted into a vector through a tokenizer and then into a latent vector through an encoder. It is mutated and used as a gene in the future. On the other hand, noise is added based on the random seed value of the image and becomes another latent vector. This is also used as a gene in the future. Based on the two vectors, the image is decoded through U-Net and VAE to generate an image. The latent vector and text vector are then submitted to U-Net [113], and the image is restored through VAE [66], a decoder, to output an image that is close to what the text means.

However, there are limitations to the Stable Diffusion model, as changing a single word or the seed value for image generation can significantly alter the output image because it introduces noise in a stochastic manner and then reconstructs the image stochastically in the learned latent space.

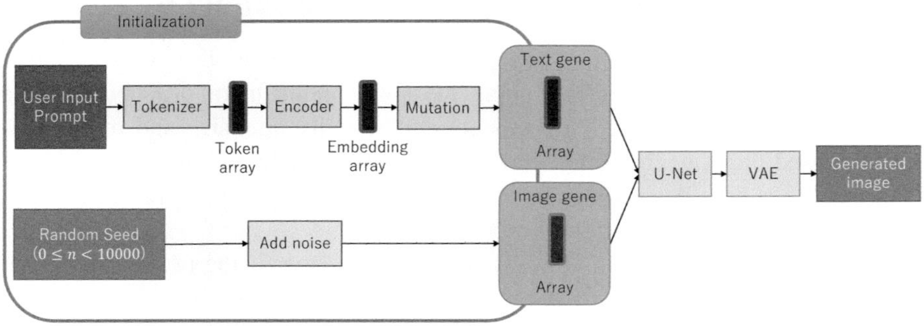


**Figure 5.1:** Overall architecture of stable diffusion using in DiffIE.

Seed 1 text A | Seed 2 text A | Seed 1 text B | Seed 2 text B

**Figure 5.2:** Images generated by stable diffusion.

Conventional methods in generative AI edit the text and change the seed value based on the generated images to create better images. However, rewriting the text is very time-consuming, but the improvement of the images is not always worth the effort. The diffusion model stochastically adds noise and then stochastically reconstructs the image in the learned latent space. Therefore, because image generation is probabilistic, there is little control over the final output. Changing a single word or changing the seed value for image generation can significantly change the output image. Figure 5.2 shows that different seed values result in significantly different images, including the color of the sand and the posture of the animals. Text A is "a cute cat sleeping on the beach." Text B is "a cute dog sleeping on the beach." In addition, when the text is changed from cat to dog, not only the type of animal but also the texture of the sand is changed, which is a significant change to the image.

By combining the diffusion model and IEC, we have proposed to explore the latent space by human evaluation. This would exclude the need to rewrite the text and to change the seed value. As new images are suggested by the selection of images, it would be easier to generate images that incorporate the user's preferences.

### 5.2.2 Diffusion Interactive Evolutionary Computation (DiffIE)

The Diffusion IEC (DiffIE) is a proposed image generation support system that combines diffusion modelling and IEC to simplify image design. We will explain the process for DiffIE below (see Fig. 5.3).

The DiffIE system uses stable diffusion as a diffusion model, which has publicly available internal programs. Tokenizer and textencoder, from the transformers library, and VAE and U-Net, from the diffusers library, are used in the inference pipeline [112, 166, 98]. The input text is first converted into a $77 \times 1$ vector using a tokenizer, and an encoder then turns it into a $77 \times 768$ latent vector, henceforth referred to as the 'text gene'. The $64 \times 64 \times 4$ noise vector in the latent space from which the image originates is referred to as the 'image gene'. With these two vectors as input, the image is generated by removing the noise using U-Net and decoding it using VAE.

For the initial individual 'text genes', the text was input by the user and mutations were added to each individual. A random number was generated for the seed value, and a $64 \times 64 \times 4$ noise vector was then generated and used as the 'image genes' of the initial individual.

The IEC algorithm used in DiffIE was adapted from DeepIE [16]. Nine images were generated based on these genes and evaluated by presenting the images to the user, who rated them as good, mutated them, and crossed them to become the next generation of images. Uniform crossover was used to incorporate features of both images used for crossover. When mutations were made, random numbers generated by the normal distribution were added, and if the number was far from the normal distribution (an absolute value of 5 or more), the absolute value would be 4.99.

The proposed image generation support system, which is named Diffusion Interactive Evolutionary Computation (DiffIE), uses a combination of diffusion modelling and interactive evolutionary computation (IEC), where candidate solutions are iteratively generated and evaluated with feedback from human evaluators, making it easier to design images. The IEC process in the image generation support system is carried out as follows:

**Initialisation** A random set of images is generated as an initial population.

**Evaluation** Each image in the population is presented to the user, who is asked to evaluate the image.

**Selection** Based on the user's evaluation, some of the images are selected to form the next generation of the population.

**Crossover** The genes of the selected images are crossed over to generate new images.

**Mutation** Genes in the generated images are randomly mutated to introduce individuals with new genes into the population.

**Repetition** The steps of evaluation, selection, crossover and mutation are repeated until a satisfactory image is generated or a stopping criterion is reached.

In Fig. 5.3, overall architecture of DiffIE is shown. As a diffusion model, stable diffusion [112], whose internal programmes are publicly available, was adopted. In addition, it allows the inference pipeline to be modified. Images generated by stable diffusion are evaluated by the user, and the genes of the images are crossed, mutated, and initialized to determine the genes of the next generation. New images are again generated based on these genes.

### 5.2.3 Can images really evolve?

The first experiment was conducted to test whether it makes sense to perform evolutionary computation by considering the latent variables of text and noise-added images as genes. Target images were prepared and the images were evolved by means of GA. Two vectors were used as genes to evolve: a noise vector from the image and a vector of latent variables from the text passed through a tokenizer. The number of individuals was 30 in total and the number of generations was 100. These values were adopted from preliminary experiments because large values for the number of generations and the number of individuals take longer to compute, small values for the number of individuals reduce the richness of the types of images generated, and small values for the number of generations do not allow evolution to finish. The top three individuals that performed well in the previous generation were left as an elite for the next generation, and five individuals were randomly selected from the previous generation to generate the next generation by crossover and mutation in a tournament strategy in which the top one of the five was used as the parent because the tournament strategy is most common method in IEC.

The mutation rate for the genes was calibrated to a value of 1 in 100, as this strikes a balance between avoiding the generation of aberrant images, which occurs at higher mutation rates, and the risk of diminishing diversity, which occurs at lower mutation rates. Random numbers following a normal distribution were added to the locations of genes that were determined to mutate, and when the absolute value exceeded 5, it was adjusted to be less than 5 [16]. Method used to determine the performance in the tournament strategy was the RGB histogram value with the target image, which is easy to calculate (see Table 5.1 for a summary of experimental set-ups).

For the first generation, a depiction expected to represent the target was considered in advance and used as input text. The genes for the first generation of images were randomly selected numbers from 1 to 10000 and assigned to each individual. The input text in the example was “a beautiful photograph of mountains, green covered hills, meadows, sunset, sky”.

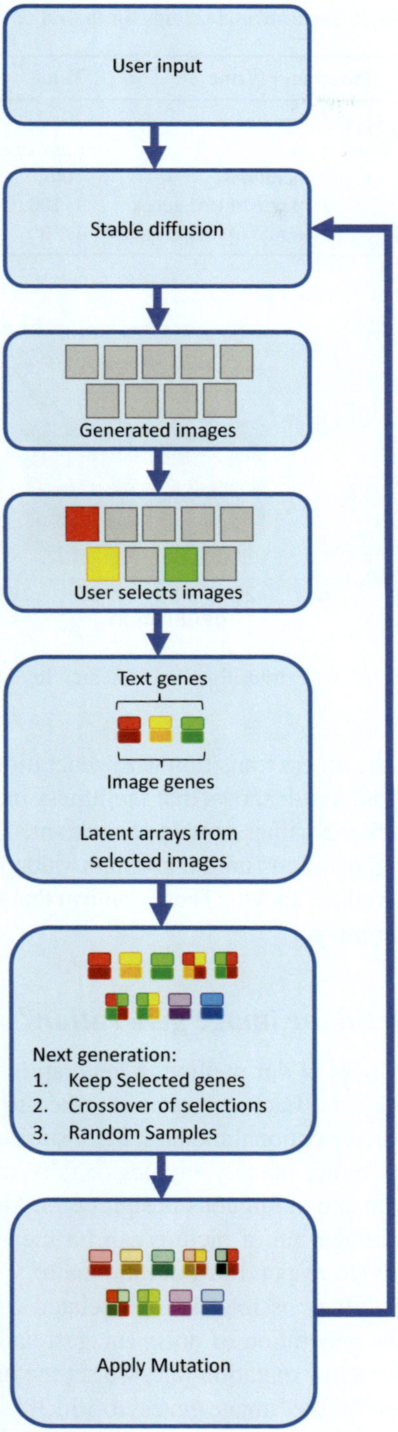


**Figure 5.3:** Overall architecture of DiffIE.

**Table 5.1:** Experimental set-ups for the validation of GA.

| Parameter Name | Value |
|---|---|
| population size | 30 |
| selection | tournament |
| #. of generations | 100 |
| mutation rate of text genes | 1/100 |
| mutation rate of image genes | 1/100 |

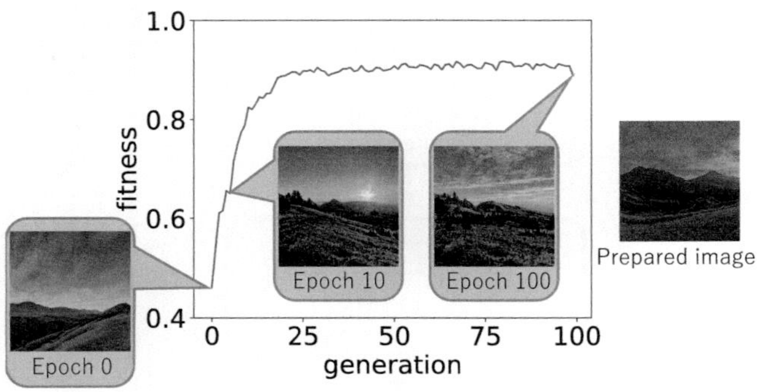


**Figure 5.4:** Fitness transition by GA search for a target image.

Figure 5.4 shows the fitness transition with generations. The average over five runs was measured. The figure shows that the fitness increased through the 20th generation, after which the values remained constant. This means that the RGB values increase with generations until a constant value is reached. Some typical images that have evolved are shown. These confirm that the GA search is steadily approaching the target image.

### 5.2.4 Is IEC useful for image generation?

To evaluate the usefulness of our method in generating images, we conducted a user survey on 36 subjects. Each subject was asked to create images by using both our program and conventional method. Each subject was assigned with two topic to create image, either nursery rhymes or fairy tales. They were adopted because they are vague and ambiguous design tasks. Thus, how well a user can convert his/her own image into a picture can be measured. Subsequently, the participants were asked to respond to a questionnaire.

Taking into account the possible risks associated with a high mutation rate, which may lead to the generation of noisy images, and the lack of diversity in images generated with a low mutation rate, we set the mutation rate for the 'text genes' to 1 in 6000 and for the 'image genes' to 1 in 300. This choice of probabilities was made based on the premise that excessively high mutation rates increase

**Table 5.2:** Experimental set-ups for DiffIE.

| **Parameter Name** | **Value** |
|---|---|
| #. of individuals | 9 |
| maximum number of generations | 10 |
| mutation rate of text genes | 1/6000 |
| mutation rate of image genes | 1/300 |

the likelihood of generating unusual images, thereby impeding the image evaluation process and potentially undermining the reputation of the DiffIE system. We set the mutation rate to be lower than that used in previous experiments evaluating the effectiveness of GA (Section 5.2.3), to balance the competing goals of image diversity and quality (see Table 5.2 for a summary of experimental set-ups).

The following process is the flow of this experiment.

**Step1** Think of prompt to go with the topic. At this point, the prompt should be as detailed as possible about what is pictured and the mood of the image.

**Step2** Input the prompt to the system.

**Step3** Choose 0 to 9 images from the 9 images created from the prompt

**Step4** Continue with **Step3** until a satisfactory image is produced. The upper limit is 10 times.

Before using the system, i.e., before moving on to the main experiment, the subjects were given practice using the GUI with a fake AI, which displays images independently of the subject's manipulation. This is to prevent results from being biased because of the loss of time due to not knowing how to use the GUI when measuring time.

Figures 5.5 and 5.6 show examples of images generated by users using DiffIE and the conventional method, respectively.

The average time required was 9.05 minutes for DiffIE and 12.6 minutes for the conventional method (see Fig. 5.7(a)). As for the one-sided Independent Samples T-test [174], the p-value was 0.00289, indicating that the time required was shorter for DiffIE at the 5% significance level. The time required here is the time from when the user enters the initial prompt and presses the "Generate Image" button to start generating images to when the user selects one image of his/her choice and presses the "Finish" button. It depended not only on the number of generations stacked by the user and the time the user spent per generation looking at and thinking about the image, but also on the free memory space of the used GPU and the communication environment because the image was sent from the server to the notebook PC.

**Figure 5.5:** Sample images generated by DiffIE.

**Figure 5.6:** Sample images generated by the conventional method.

In contrast, the average number of stacked generations was 6.94 for DiffIE and 7.00 for the conventional method (see Fig. 5.7(b)). The p-value was 0.465, not significantly different.

Next, the following questionnaire was conducted to determine user satisfaction. The first questions were included in the Likert scale survey. The participants answered each item on a scale of 1 (strongly disagree) to 5 (strongly agree).

1. The image generated was satisfactory.

2. It supplemented imagination.

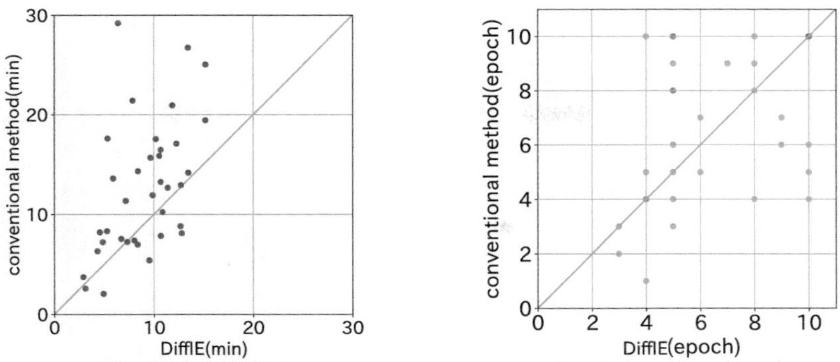


(a) Required work time.

(b) Required number of epochs.

**Figure 5.7:** Time and epoch required for each method. Each point in the figure corresponds to each subject (time or epochs spent in each method).

3. It was easy to change design.

4. It was useful to create the image as you want.

In addition, the following free-response questions were asked:

1. Which system was easier to use?

2. Which system do you prefer and why?

The Wilcoxon signed-rank test [110] was used to compare the responses obtained from the Likert scale. When participants were asked to rate their satisfaction with the generated images on a 5-point scale, the results yielded a non-significant p-value of 0.103 (see Fig. 5.8(a)), indicating that no significant differences were found between the DiffIE and conventional methods. Likewise, when participants were asked to rate on a 5-point scale the extent to which the system assisted their imagination in generating the images, the results yielded a non-significant p-value of 0.587 (Fig. 5.8(b)), indicating that no significant differences were found between the DiffIE and conventional methods. When participants were asked to rate the ease of changing the image design on a 5-point scale, the results also yielded a non-significant p-value of 0.617 (Fig. 5.9(a)). But the p-value was 0.0215 (Fig. 5.9(b)), indicating that DiffIE was significantly more useful than the conventional method at a 5% level of significance.

As for the user preference the one-sample one-tailed T-test [87] was conducted with the null hypothesis that the mean is less than 3. When asked which would be easier to use, 22 respondents favored DiffIE, 13 favored the traditional method, and 1 favored neither. The p-value was 0.0431, indicating that DiffIE was rated higher at the 5% level of significance (see Fig 5.10(a)). When asked which would be better for future designs, 21 respondents favored DiffIE, 14 favored the traditional method, and 1 favored neither. In this case, the p-value

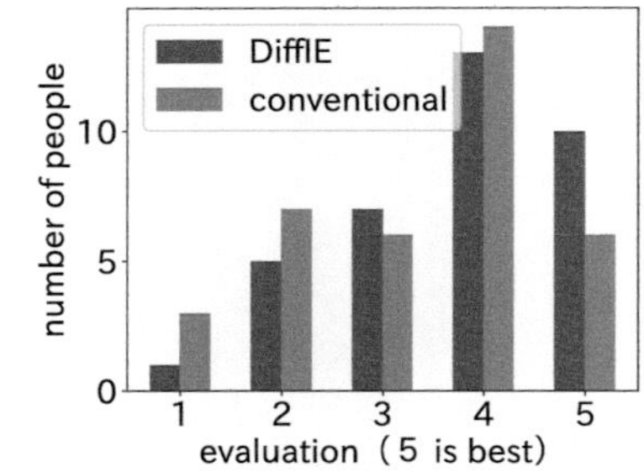


(a) The satisfaction with the generated images.

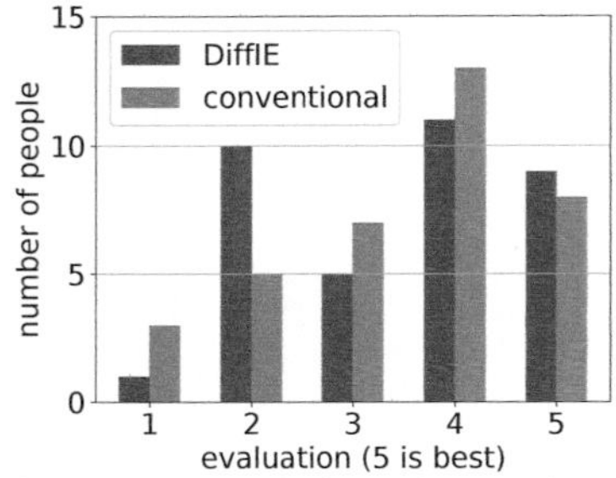


(b) How well each method complements the user's imagination.

**Figure 5.8:** User ratings.

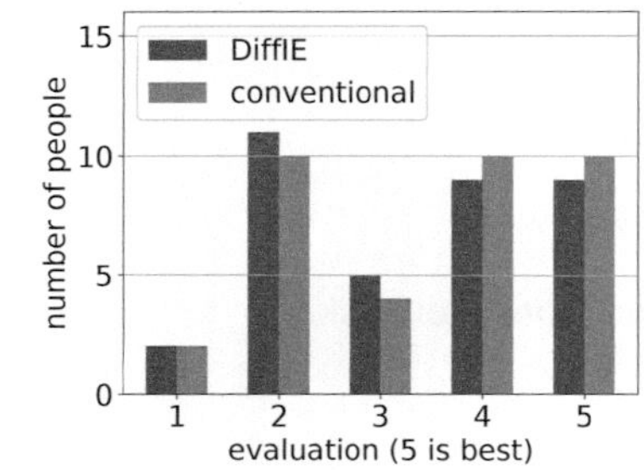


(a) The ease of design changes for each method.

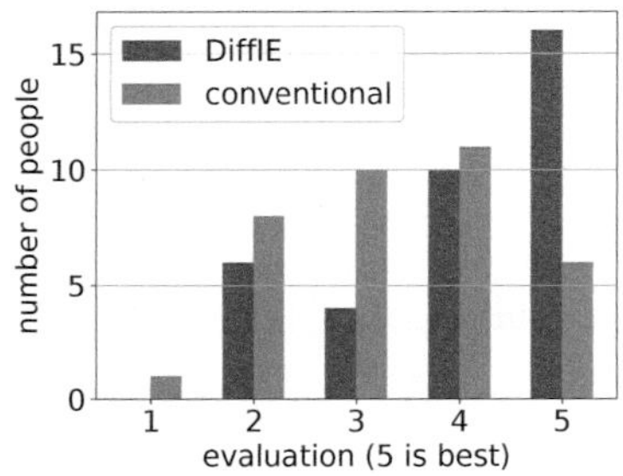


(b) The usefulness to generate a desired image for each method.

**Figure 5.9:** User evaluation.

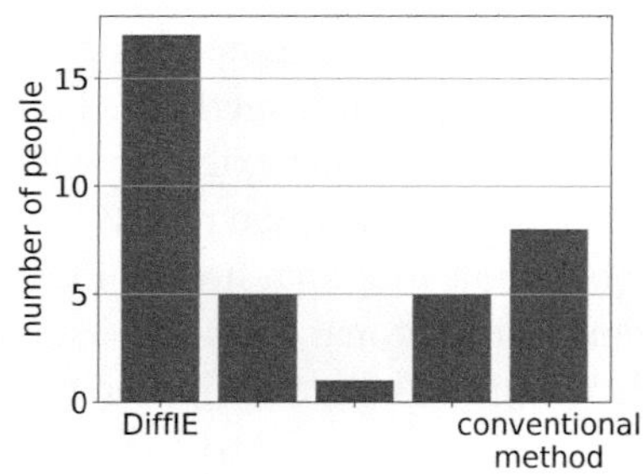


(a) Which method was easier to use.

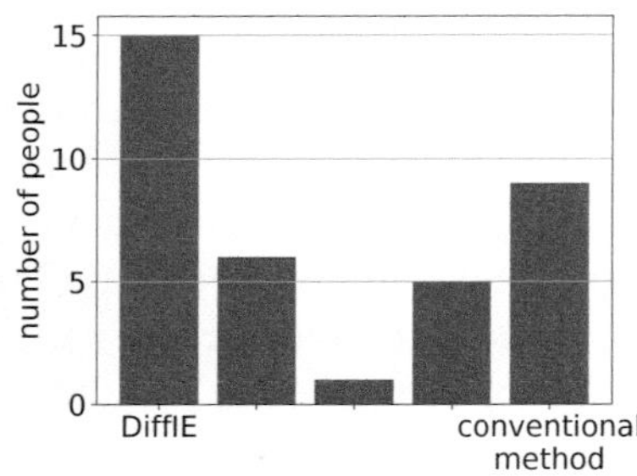


(b) Which method was preferred.

**Figure 5.10:** User preference.

was 0.107, and no significant difference existed at the 5% level of significance (Fig 5.10(b)).

Supporters of DiffIE said that it allows for fine-tuning, that it is easy to reflect images in the brain because "images" are selected, and that selecting images is easier than updating the text. On the other hand, those who supported the conventional method said that they were dissatisfied with DiffIE because the initial text could not be changed, and that DiffIE did not expand their ideas because they were presented with similar images. Those who were not in favor of either method said that DiffIE, which allows fine-tuning, and the conventional method, which allows rough changes, have their own advantages, and cannot be compared in general.

### 5.2.5 Discussion

This section described investigated the effectiveness of DiffIE as a tool for generating images, with a particular focus on its impact on the image creation process, user satisfaction, compensation, flexibility, and preferences. Results of user experiments revealed that the use of DiffIE resulted in a reduction in the time required for image creation, and also facilitated the image creation process, suggesting its efficacy as an image generation tool. The ease of image creation was found to be enhanced as a result of the reduction in the need for text updating and the availability of intuitive image selection choices.

However, no significant differences were observed between the use of DiffIE and conventional methods with respect to satisfaction, compensation, flexibility, and user preferences. It is hypothesized that some users may have felt that DiffIE compensated for their lack of imagination by enabling them to see more images. User preferences were found to be divided between the benefits of DiffIE, which facilitates fine adjustments, and the advantages of conventional methods, which facilitate major changes.

The advantage of DiffIE lies in its ability to reduce the burden of image design by proposing a new image with each iteration. For this purpose, we are aiming at the development of an algorithm that can evolve images while allowing for the modification of the initial prompt. It is worth noting that DiffIE's performance in accommodating major changes was found to be limited, and thus, its effectiveness could be enhanced by expanding the search range. These findings highlight the potential of IEC as an effective tool for improving the image generation process while also highlighting the need for further refinement of the DiffIE system.

## 5.3 Co-design framework of robot morphology by means of neuro-evolution

Creating intelligent agents capable of autonomous operation in complex environments is a significant focus in artificial life research. Soft robots, composed of pliable materials resembling living organisms, hold potential industry impact due to enhanced adaptability compared to rigid robots, facilitating safe collaboration with humans [114]. However, designing soft robots is more challenging than rigid robots, as predicting the dynamics of their behavior and determining adaptability to an environment is hindered until controller optimization [151].

Natural life concurrently optimized morphology and brain over time. Co-design, applying this evolutionary process, automatically optimizes a robot's structure and controller [10]. Existing studies proposed co-design algorithms, combining structure optimization and control optimization, demonstrating effectiveness in soft robot design through simulation experiments [20, 23, 24, 150, 10]. However, designing soft robots for complex tasks remains challenging [10].

In this section, we explain a lamarckian co-design method for obtaining more performant soft robots in less computational time by inheriting controllers optimized through DRL. We train newly generated robots in each generation using Proximal Policy Optimization (PPO), a state-of-the-art DRL (Deep Reinforcement Learning) method. Consequently, robots exhibiting superior performance during the training phase function as parent individuals, generating offspring that receive both the controller and the physical structure. To address the challenge of sensor and actuator differences between parent and child robots, we utilize a Transfer Learning (TL) approach. We show the effectiveness of our approach by applying to EvoGym framework.

Evolution Gym (Evogym) [10] serves as a benchmark for soft robot co-design algorithms, enabling robots to tackle various tasks. Bhatia et al. [10] tested existing algorithms on Evogym, finding the most effective method combined GA for structure optimization and DRL for controller optimization. Nevertheless, this approach faces challenges related to extensive training time and difficulties in handling complex tasks [10]. We attribute these challenges to the necessity of training each robot individually from the start, arising from the differences in sensors and actuators among individuals. This variability makes it challenging to inherit controllers from parents to children. Although there exist strategies for inheriting controllers enhanced via GA [136], there remains a lack of well-established procedures for inheriting controllers refined using DRL.

## 5.3.1 Robot co-design and Evolution Gym

The pioneering attempt to co-optimize the structure and controller can be traced back to Sims' work [122]. Sims designed virtual creatures with rigid parts for structures, representing their controller as Artificial Neural Networks (ANNs) in the form of directed graphs. Employing GA with task achievement as the fitness criteria, Sims successfully evolved virtual creatures capable of walking, swimming, or jumping. Subsequent to Sims' work, co-design studies for rigid robots have gained prominence [156, 46, 96, 177, 157]. However, rigid robots faced challenges in tackling complex tasks due to their limited degrees of freedom [10].

Consequently, attention shifted towards the co-design of soft robots, constructed from flexible materials mimicking real organisms. Soft robots, a novel category, consist of flexible materials like polymers, rubber, and silicone [76].

Compared to traditional rigid robots, soft robots offer advantages such as impact absorption during collisions, facilitating safe collaboration with humans. Their high degree of freedom, attributed to the ability to continuously alter their structure shape, enables effective adaptation to diverse environments [114].

Several attempts have been made to co-design soft robots in simulations [20, 23, 24, 150]; however, these studies exhibit some limitations. In these methods, robot control is often constrained to periodic and simplistic patterns, and the assigned tasks are frequently limited to simple activities such as walking on flat surfaces. Consequently, adapting soft robots to complex environments becomes challenging. Additionally, the difficulty in comparing algorithm performance arises from each study conducting experiments in its own uniquely developed environment [10].

To tackle these issues, Bhatia et al. [10] introduced Evolution Gym (Evogym), a simulation environment designed for co-designing soft robots. In Evogym, soft robots are constructed in a 2D space by assembling four types of voxels made from soft materials. These soft robots possess the ability to deform their structure based on environmental observations, enabling them to execute various tasks such as locomotion, object manipulation, climbing, and more. In this section, we delve into the specifics of the simulation environment provided by Evogym.

The robots in Evogym consist of four voxel types: black rigid voxels, gray soft voxels, light blue horizontal actuators, and orange vertical actuators. An example of a robot's structure is presented in Fig. 5.11(a).

At each time step, the robot takes actions by extending or contracting actuators based on observations from sensors. The sensors include:

- **Position Sensor**: Located at voxel vertices (indicated by red circles in Fig. 5.11(b)), this sensor provides the relative position from the robot's center of mass to itself.
- **Velocity Sensor**: This sensor offers the velocity of the robot's center of mass.
- **Rotation Sensor**: This sensor supplies the rotation angle of the robot.
- **Object Sensor**: In tasks involving object manipulation, this sensor provides two pieces of information: 1) the relative position of the target object's center of mass from the robot's center of mass, and 2) the velocity of the target object.

The availability of sensors varies based on the task, but the position sensor is consistently present for all tasks. Since position sensors are positioned at the vertices of the voxels forming the robot, their quantity depends on the robot's structure, while the robot is equipped with only one sensor each for velocity, rotation, and object. The blue-outlined colored voxels in Fig. 5.11(c) serve as actuators. Each actuator receives a real-valued input within the range of $[0.6, 1.6]$ from the controller and adjusts its extension or contraction based on the input magnitude.

Since Evogym is a relatively recent development, there are still few research examples utilizing Evogym. Next section introduces some prior studies that serve as baseline algorithms for comparison with our research.

### 5.3.2 Related works

Bhatia et al. [10], the creators of Evogym, conducted experiments on Evogym using three structure optimization algorithms: GA, CPPN-NEAT (Compositional Pattern Producing Networks - NeuroEvolution of Augmenting Topologies) [128, 127], and Bayesian optimization, each combined with PPO [118]. Through these experiments, they found that the most effective approach for structure optimization is GA.

In this GA-based approach, the robot's genome is represented by an integer matrix of the same size as the robot's structure. In this matrix, the element 0 indicates the absence of a voxel at that location, while elements 1 through 4 correspond, respectively, to a rigid voxel, soft voxel, vertical actuator, and horizontal actuator. For example, the genome of the robot in Fig. 5.11 is represented as follows.

$$\begin{bmatrix} 0 & 0 & 0 & 2 & 3 \\ 2 & 0 & 4 & 4 & 4 \\ 1 & 3 & 2 & 0 & 1 \\ 1 & 1 & 1 & 3 & 4 \\ 0 & 3 & 0 & 2 & 0 \end{bmatrix}$$

The algorithm operates as follows: In each generation, newly born robot individuals within the population undergo PPO learning from scratch, and the results of this learning are assigned as fitness. After learning, only elite individuals are preserved while the rest are eliminated. The eliminated individuals are then replaced by generating new individuals through mutation from the elite individuals. Mutation occurs with a certain probability for each element of the genome matrix, and the mutated elements are assigned random values from 0 to 4.

CPPN-NEAT has been employed for robotics (see Section 5.1). However, the approach of encoding robot controllers using CPPN involves optimization solely through GA, without utilizing gradient-based learning. In a comparative experiment conducted by Saito et al. [115] using evogym, three different optimization methods, namely CPPN-NEAT, HyperNEAT, and PPO, were employed for controller optimization. The findings revealed that PPO demonstrated superior performance in the majority of tasks.

RL (Reinforcement Learning) serves as a valuable tool for training robots to execute complex control tasks without relying on training data. In RL, an agent receives a state $s_t \in \mathcal{S}$ from the environment at each time step $t$. Subsequently, the agent decides on an action $a_t \in \mathscr{A}$, and at the next time step $t+1$, it receives a reward $r_{t+1}$ based on the state-action pair $(s_t, a_t)$. Here, $\mathscr{S}$ and $\mathscr{A}$ represent the state and action spaces, respectively, indicating the sets of possible states and actions available to the agent. The agent's goal is to acquire an optimal policy $\pi^*(s)$—a rule for action selection based on states—that maximizes the cumulative reward, given by the sum $\sum_{t=0}^{\infty} \gamma^t r_t$, where $\gamma \in [0, 1]$ is a constant known as the discount factor. The value function $V^\pi(s_t)$ estimates the expected cumulative reward at time $\sum_{k=0}^{\infty} \gamma^k r_{t+k+1}$ for a given state $s_t$. The agent learns the optimal policy by iteratively estimating the value function $V^\pi(s)$ through interactions with the environment based on the policy $\pi(s)$ and adjusting the policy $\pi(s)$ using the value function $V^\pi(s)$ [133].

Transfer Reinforcement Learning (TRL) is a method where knowledge acquired by one agent through RL aids in the learning of other agents. The primary objectives of TRL are as follows [75]:

- **Learning Speed Improvement**: Achieving a reduction in the number of iterations required to learn the optimal policy.
- **Asymptotic Improvement**: Attaining a more optimal policy compared to scenarios without knowledge transfer.
- **Jumpstart Improvement**: Obtaining an improved policy in the initial stages of learning compared to scenarios without knowledge transfer.

Taylor et al. [137] proposed a method called Behavior Transfer for TRL between two RL tasks $T_1$ (state space $\mathscr{S}_1$, action space $\mathscr{A}_1$) and $T_2$ (state space $\mathscr{S}_2$, action space $\mathscr{A}_2$), where generally $\mathscr{S}1 \neq \mathscr{S}2$ and $\mathscr{A}1 \neq \mathscr{A}2$. In Behavior Transfer, the policy $\pi_{(1,final)}$ obtained from RL in $T_1$ is transformed by a manually designed policy transform function $\rho$. This transformed policy $\rho(\pi_{(1,final)})$ is then used to initialize the policy $\pi_2$ for $T_2$ (i.e., $\pi_{(2,initial)} = \rho(\pi_{(1,final)})$).

### 5.3.3 LAMACODA (LAMarckian CO-design Algorithm)

As outlined in previous sections, existing co-design algorithms for soft robots fall into two categories: those optimizing the robot's controller using DRL without inheriting the controller [10], and those inheriting the controller without utilizing DRL for optimization [20, 23, 24, 136]. In this section, we explain a Lamarckian co-design algorithm (see Section 1.1.1), which is called LAMACODA (LAMarckian CO-design Algorithm), for soft robots wherein the optimized controller obtained through DRL is inherited from parents to children. The aim is to reduce co-design time and enable the robot to perform complex tasks that existing methods cannot achieve. Additionally, to demonstrate the effectiveness of inheriting learned controllers when generating robot structures by crossing the genomes of two robot structures (a feature not used in Bhatia et al.'s method [10]), we propose a simple method for crossing structure genomes.

**Algorithm 5** LAMACODA (LAMarckian CO-design Algorithm).

**Input:** population size $p$, max generations $n$

**for** $k = 1$ to $p$ **do**<br>
    **initialize** structure $S_k$ and controller $C_k$.<br>
**end for**<br>
**for** $generation = 1$ to $n$ **do**<br>
    **for** $k = 1$ to $p$ **do**<br>
        **if** $S_k$ is a newborn **then**<br>
            **train** $C_k$ for $S_k$.<br>
            **let** $f_k$ be the best score achieved by $(S_k, C_k)$.<br>
        **end if**<br>
    **end for**<br>
    **perform natural selection**<br>
    **for** $k = 1$ to $p$ **do**<br>
        **if** $k$ has died **then**<br>
            **mutate** or **crossover** elites' structure to configure $S_k$.<br>
            **inherit** parents' controller and configure $C_k$.<br>
        **end if**<br>
    **end for**<br>
**end for**<br>
$l = \arg\max_{k} (f_k)$ ↵ $S_l, C_l$

The LAMACODA algorithm is presented in **Algorithm 5**. The processes of mutation and natural selection followed the same method employed by Bhatia et al., as described in Section 5.3.2.

To encode the structure using genomes, we employed the same method utilizing integer matrices as introduced in Section 5.3.2. For genome crossover between two parents, we initially cut the two genomes along a randomly selected line (available cutting lines are depicted in Fig. 5.12), then merged them to produce two new genomes. The offspring's genome was randomly selected from these two new genomes. An illustrative example of crossover is presented in Fig. 5.13. The child is configured by crossing over between parent1 and parent2.

We performed controller inheritance using TRL. The actor-network, governing control in PPO, comprises a fully connected input layer, two intermediate layers, and an output layer. The input layer's node count corresponds to the robot's state space dimension, the output layer's node count corresponds to the action space dimension, and both intermediate layers have a fixed node count of 64. Given that the state and action spaces depend on the robot structures, the actor-network structures differ between parents and children. Therefore, to inherit learned controllers from parents to children, an appropriate policy trans-

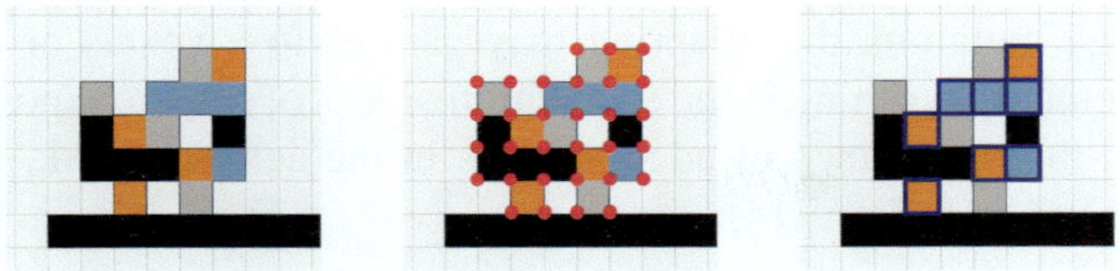

(a) structure (b) position sensors (c) actuators

**Figure 5.11:** An example of soft robot in Evogym.

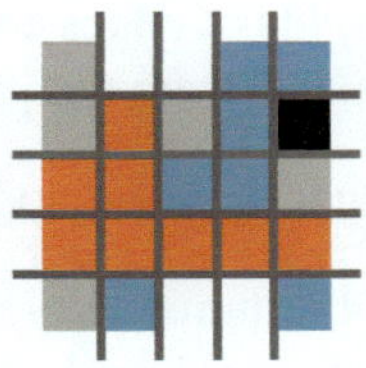

**Figure 5.12:** Available cutting lines for a robot with a $5 \times 5$ shape.

(a) parent1 (b) parent2 (c) child

**Figure 5.13:** An example of crossover.

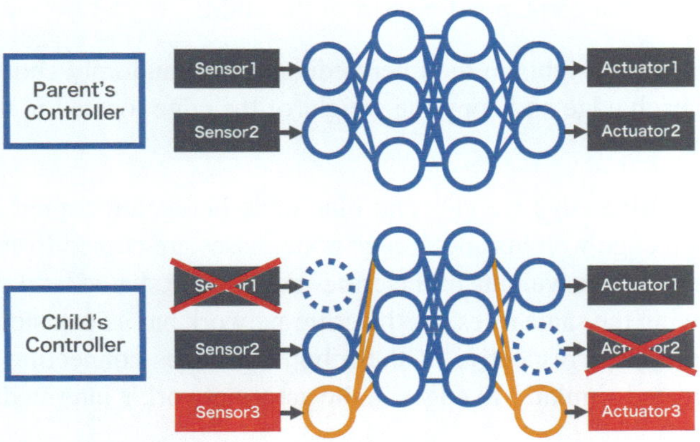


**Figure 5.14:** An example of controller inheritance through mutation.

form function is necessary. Behavior transfer with LAMACODA was achieved by copying the weights of connections in the actor-network.

Through mutation, the offspring may gain or lose voxels or change voxel types, potentially resulting in the acquisition or loss of actuators/position sensors. To govern the inheritance of edge weights in the actor-network, the following rules were applied (Fig. 5.14):

- Initialize weights of edges connected to nodes linked to newly acquired actuators (or position sensors) through mutation.
- If the child has an actuator (or position sensor) at the same position as the parent, copy the weights of edges connected to the node linked to that actuator (or position sensor) from the parent.
- Copy the weights of edges between nodes in the intermediate layers.

In Fig. 5.14, the lower part of the figure depicts the controller of a child robot, which, through mutation, lost Sensor1 and Actuator2 but gained Sensor3 and Actuator3. Blue edge weights and blue node biases are copied from the parent, while orange edge weights and orange node biases are initialized.

Through crossover, the child inherits actuators and position sensors from both parents. The inheritance of edge weights in the actor-network follows these rules (see Fig. 5.15):

- If an actuator (or position sensor) originates from either parent, copy the weights of edges connected to the node linked to the actuator (or position sensor) from the parent.
- If a child's position sensor is located where both parents have one, randomly choose one parent and copy the weights of edges connected to the node to the corresponding part of the child.
- For edge weights in the intermediate layer, randomly choose one parent for each edge and copy the weight of the edge located in the same position.

In Fig. 5.15, Blue edge weights and blue node biases are copied from Parent1, while orange edge weights and orange node biases are copied from Parent2.

Furthermore, the weights of the critic-network are shared between parent and child following the same rules. As the critic-network has a consistent output layer size of one node, irrespective of the robot, the edges connecting to the output layer are treated similarly to edges in the actor-network's intermediate layer.

### 5.3.4 Experimental set-ups

We conducted experiments on Evogym [10] involving various tasks, comparing three algorithms: (i) LAMACODA with controller inheritance and crossover, (ii) LAMACODA with controller inheritance but without crossover, and (iii) the

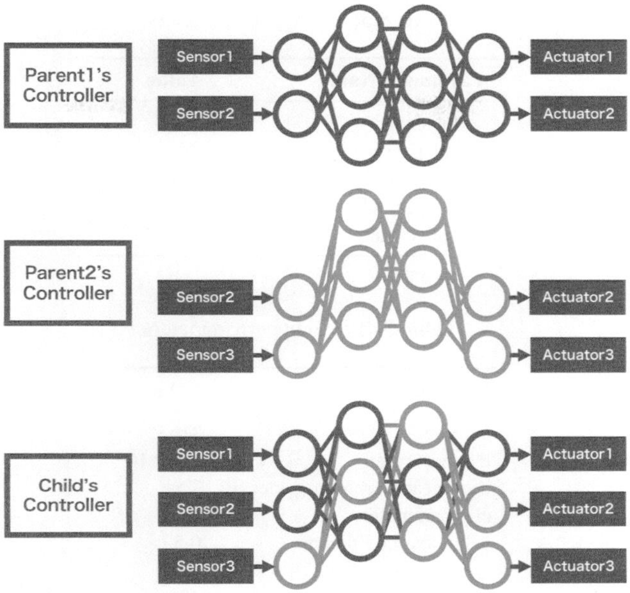


**Figure 5.15:** An example of controller inheritance through crossover.

existing method without crossover or control transfer. In cases of (ii) and (iii), robots reproduced only through mutation. In contrast, half of the individuals experienced mutation in (i), and the remaining half underwent crossover for reproduction. For crossover, two randomly selected distinct elite individuals were chosen as parents.

The hyperparameters for the genetic algorithm are presented in Table 5.3, while the key hyperparameters for PPO are shown in Table 5.4. Survival rate was linearly reduced from 0.6 to 0.0. However, irrespective of how low the survival rate became, a minimum of two individuals were retained as elite individuals. Probability of gene mutation indicates the probability that each voxel of the robot will change into a different type of voxel during the mutation process. Crossover rate represents the proportion of robots within the population that are born through crossover. The number of robots to be trained (synonymous with the number of generations) was set for each task based on the reference [10]. The implementation of PPO utilized the version developed by Kostrikov [73]. For a fair comparison, all parameters, except for the presence of controller inheritance and crossover, were set to be identical across the three algorithms.

We employed six tasks for the experiments, outlined as follows. Refer to [10] for task details.

- **BridgeWalker-v0:** Easy task of traversing an unstable terrain made of soft material, with reward based on horizontal distance moved. A total of 250 robots are trained for this task.

**Table 5.3:** GA hyperparameters.

| Parameter Name | Value |
|---|---|
| robot size | $5 \times 5$ |
| population size | 25 |
| survival rate | $0.6 \to 0.0$ |
| probability of gene mutation | 0.1 |
| crossover rate | 0.5 |

**Table 5.4:** PPO hyperparameters.

| Parameter Name | Value |
|---|---|
| use gae | True |
| learning rate | $2.5 \times 10^{-4}$ |
| use linear learning rate decay | True |
| clip parameter | 0.1 |
| value loss coefficient | 0.5 |
| entropy coefficient | 0.01 |
| num steps | 128 |
| num processes | 4 |
| evaluation interval | 50 |
| train iters | 1000 |

- **BidirectionalWalker-v0:** Medium-difficulty task of changing movement direction towards a randomly switching goal point, with reward based on distance moved towards the goal. A total of 750 robots are trained for this task.
- **ObstacleTraverser-v1:** High-difficulty task of traversing an uneven path, with fitness determined by total horizontal distance moved. A total of 750 robots are trained for this task.
- **Carrier-v0:** Easy task of transporting objects on a flat path, with fitness determined by total distance the robot transports the target object, penalized if the object is dropped. A total of 500 robots are trained for this task.
- **Carrier-v1:** High-difficulty task of transporting objects on an uneven path. A total of 330 robots are trained for this task.
- **Pusher-v0:** Easy task of pushing an object forward on a flat path, with fitness determined by total distance the robot pushes the target object. A total of 750 robots are trained for this task.

### 5.3.5 Results and discussion

The best performance curves of three algorithms across six selected tasks are presented in Fig. 5.16. We plotted the fitness of the best-performing robot in each generation. All curves represent the average over three different runs, and the variance is indicated by shaded regions.

The following improvements were observed using controller inheritance:

- A reduction in the number of individuals required for robots to attain the fitness of the best-performing robots obtained by the existing method. The average reduction was 41.8% without crossover and 37.3% with crossover.
- An average improvement in the fitness of the final evolved robots of 11.2% without crossover and 11.8% with crossover.
- Excluding tasks BridgeWalker-v0 and Carrier-v1, which were already completely accomplished by the existing method, improvements of 16.7% and 28.1%, respectively, were observed.

Additionally, crossover contributed to additional performance enhancement in ObstacleTraverser-v1. While there was no significant difference in optimization speed with controller inheritance in Carrier-v0, the use of crossover expedited the optimization process.

Learning curves depicting the trajectory of the robot's cumulative reward for each method are presented in Fig. 5.17. We plotted the trajectory of the robot's cumulative reward. All the curves are averaged over all of the robots in 3 different random seeds, and the variance is indicated by shaded regions. Note that the shaded regions may appear small due to the small magnitude of the error.

In existing methods, robots were scarcely able to accomplish tasks before learning. In contrast, LAMACODA enabled robots to exhibit some level of task achievement even before learning. Moreover, at the end of the learning process, the performance of robots in LAMACODA surpassed that of the existing method.

We posit that the superiority of LAMACODA of (i) and (ii) over the existing method (iii) can be attributed to the following reasons:

- **Improvement in Learning**: The inheritance of the control system significantly reduced the learning cost for robots, facilitating the acquisition of more complex behaviors.
- **Unimpeded Structure Evolution**: The inheritance of the control system did not impede the morphological evolution of the robots.

The result indicated in Fig. 5.17 underscores that inheriting controller, leading to jump-start and asymptotic improvement, contributes to the efficiency of learning. In certain instances, LAMACODA led to robots demonstrating distinct

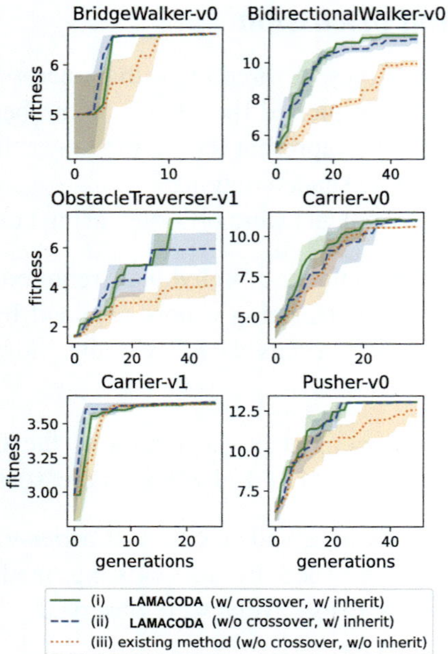


**Figure 5.16:** The impact of inheritance and crossover on evolution.

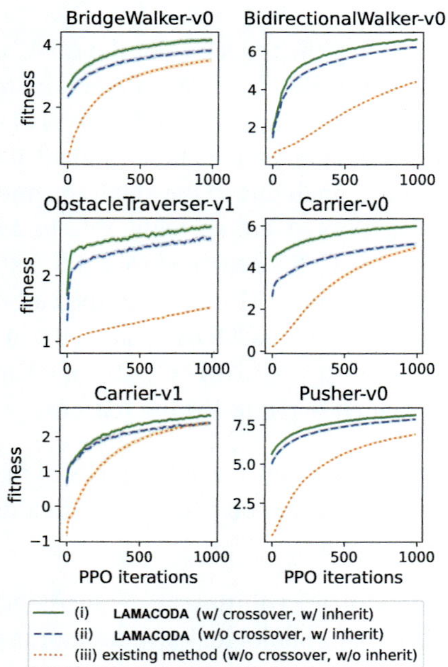


**Figure 5.17:** The impact of inheritance and crossover on PPO learning.

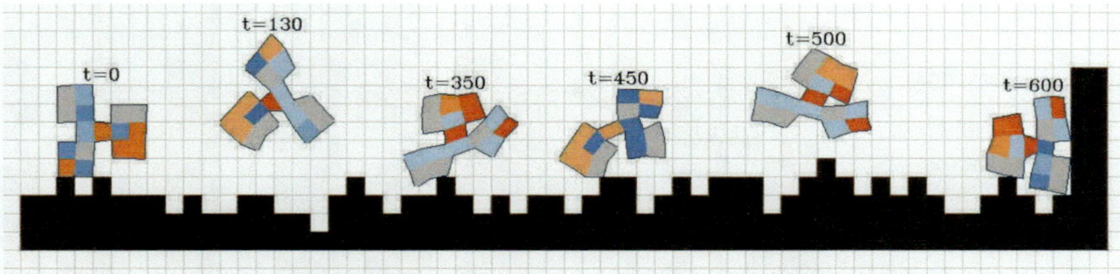


(a) LAMACODA (w/o crossover, w/inherit)

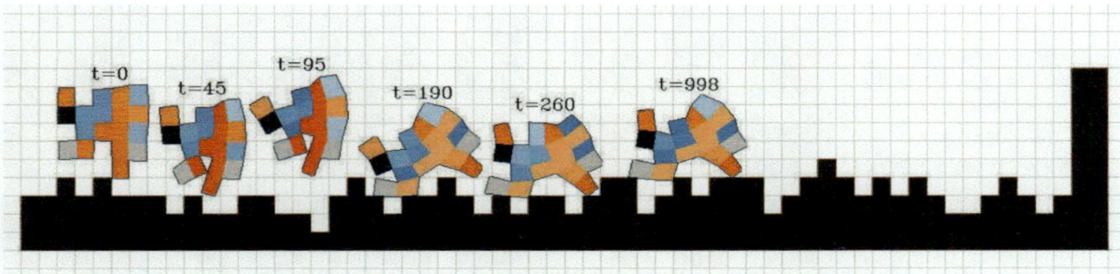


(b) Existing method (w/o crossover, w/o inherit)

**Figure 5.18:** Comparison of behaviors of the best robots.

control strategies in comparison to the existing method. A case in point is illustrated in Fig. 5.18. We illustrated the behaviors of the best individuals obtained in a single run in ObstacleTraverser-v1 for both LAMACODA without crossover and the existing method. The best robot from LAMACODA rotates its structure to avoid obstacles and traverses more effectively compared to the best robot from the existing method, which jumps high but comes to a stop midway. This trend was consistently observed across all runs of the ObstacleTraverser-v1 task.

A major concern in establishing our methodology was the potential adverse impact of inheriting controllers on the evolution of body structures. In LAMACODA, the differences in the body structures of parent and child robots, caused by mutations, are determined randomly. Consequently, a child robot acquiring or losing voxels, changes the number of initialized parameters but also diverges in optimal control strategies from the parent. This suggests that inheritance may be more beneficial for small structural changes. However, it was observed that regardless of the method used, the robots' structures converged to similar structures or structures with similar functions. Evolved robot structures in various tasks are compared in Fig. 5.19. For each method and each run, the structures of robots that achieved the most optimal fitness in each trial were illustrated.

The evolved structures for robots are summarized in Table 5.5.

The incorporation of crossover, along with controller inheritance, contributed to enhanced co-design efficiency and the ultimate performance of evolved robots in Carrier-v0 and ObstacleTraverser-v1 tasks. We attribute this improvement to the combining structures from distinct elite individuals, leading to the development of innovative structures.

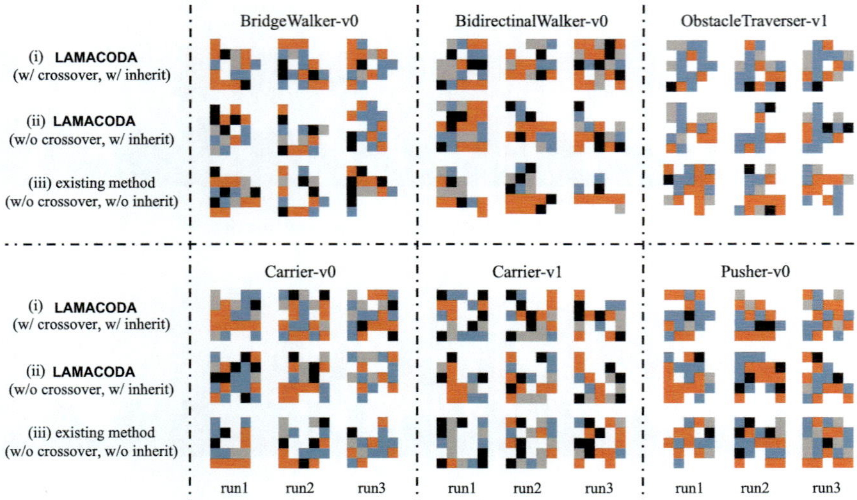


**Figure 5.19:** Comparison of evolved bodies.

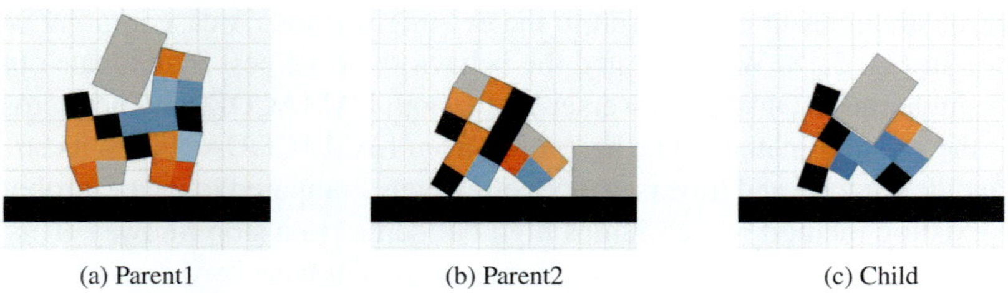


**Figure 5.20:** Example crossover in Carrier-v0 experiment.

Figure 5.20 shows an example crossover in Carrier-v0 experiment. A child (fitness:7.97) is generated by combining parent1 (fitness:4.52) with parent2 (fitness:4.77). In the figure, Parent1 possesses an upper structure designed for holding objects but is equipped with only horizontal actuators in the lower part, making it challenging to move forward. On the other hand, Parent2 has a lower structure with vertical actuators, and its contour is well-suited for maintaining a forward-leaning posture, enabling strong forward propulsion. However, its upper structure is not designed for holding objects. The child inherits the upper structure responsible for holding objects from Parent1 and the lower structure for propulsion from Parent2, resulting in a configuration well-suited for transporting objects over a distance.

**Table 5.5:** Evolved structures for robots.

| **Task** | **Detail description** |
|---|---|
| BridgeWalker-v0 | Structures with many voxels in the lower part to increase ground friction or structures without voxels in the lower right part to facilitate a forward-leaning posture. |
| ObstacleTraverser-v1 | Structures with multiple actuators connected in series, enabling robots to strongly push off the ground for leaping over obstacles. |
| Carrier-v0 | Structures resembling an upper recess for holding objects and a lower structure suitable for walking, either in the form of animal-like legs or by concentrating actuators in the lower part. |
| Carrier-v1 | Structures with an upper structure acting as a container and a lower structure with a flat contact surface for stable object transportation on stair-like terrain. |
| Pusher-v0 | Structures resembling two legs of an animal or a forward protrusion, suitable for kicking the object far away. |

## 5.4 Rule Optimization via Large Language Models

As discussed in Section 5.4 (1) **Transferability** and (2) **Safety** are important in real-world applications of LLM. This section explains a novel method called Rule Optimization via Large Language Model for Imitation Learning (ROIL). ROIL is designed for optimizing human-readable rules in a Natural Language Learning Classifier System (NL-LCS) [53]. ROIL accomplishes transferability and safety by learning from human demonstrations, which is inspired by imitation learning of the reinforcement learning field. An overview of the method is shown in Fig. 5.21. Due to its simplicity, the use of ROIL in NL-LCS serves as foundational research that can be built upon for further investigation in this area. ROIL is an effective framework for optimizing human-readable rules in a NL-LCS using LLMs, offering both transferability and safety over existing methods.

### *5.4.1 Learning classifier systems*

Classifier system (CS) is a typical example where a GA is applied to machine learning, and has been studied by many researchers. A CS uses a GA to enhance and generate rule-based knowledge in achieving (1) knowledge acquisition in complex systems and (2) generation of appropriate output.

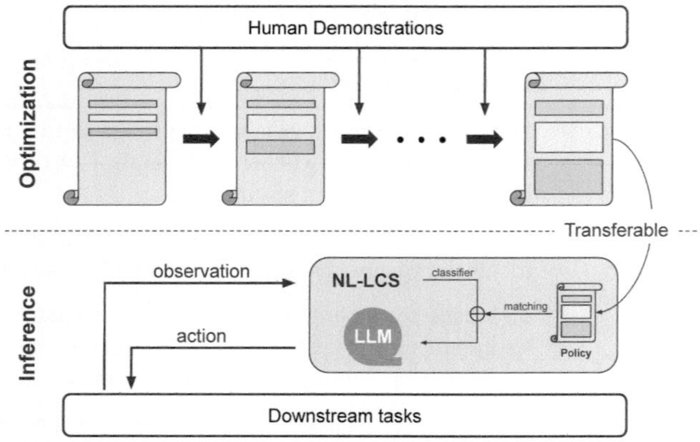


**Figure 5.21:** Overview of ROIL highlighting the two phases: rule optimization from human demonstration and action inference using the learned transferable rules.

Machine learning using a GA is called genetic-based machine learning (GBMl). The cognitive system level–1 (CS1) by Holland[1] is a famous early example of GBML [53]. Holland and his colleagues used this system to learn how to search in maze problems. Stephen Smith later developed a learning system called LS1 [124], which was applied to maze searches and poker game strategy, and whose effectiveness has been empirically demonstrated.

GBML led to two approaches, that is, the Michigan and Pittsburgh approaches. The difference between these approaches is whether the number of rule sets is just one or more than one. The Michigan approach is based on CS1, where one rule is considered as one individual, and is the main type of CSs. The details of these two approaches will be explained later.

GBML differs from optimization, which searches for a solution close to the optimum solution. Instead, machine learning generates new structures and obtains a coordinated set of rules while incorporating given information. Therefore, GBML must consider the following.

1. New rules are continuously generated, and good rules remain while bad rules are discarded.

2. Good rules generated during the learning process are not destroyed in later learning.

3. The number of rules is not limited, and retaining all necessary rules is possible.

4. Similar rules are sorted out to generate a rule set with little redundancy.

---

1  John Henry Holland (1929–2015): American scientist. He is known for his pioneering work in complex systems and nonlinear science. In 1975, he wrote the seminal book "Adaptation in Natural and Artificial Systems" and became the father of Genetic Algorithms. He also devised the "Schema theorem" for the purpose of GA theoretical analysis.

CSs use the following symbol in learning to achieve a creative rule generation mechanism:

```
If <<condition>> then <<action>>
```

This is the same as the production rule often used in expert systems. The rules in CSs consist of a condition that can be expressed with a string of 0, 1, and # (don't care) and an action that can be expressed with a string of 0 and 1. Here, # is a string that matches both 0 and 1. Providing an external message from the environment to CS results in simultaneous booting of many rules in the system that gives a corresponding output. In other words, the action part of rules is executed if the input message matches the condition part.

CS consists of the following three elements:

- **Performance Component**: It is a kind of production system that searches for a classifier which matches the message, and finds the output of the CS system.
- **Reinforcement Component**: Reinforcement learning of classifiers is performed.
- **Discovery Component**: Discovery learning of classifiers is carried out by means of evolutionary computation.

The execution part finds a matching classifier for the external input and executes its consequent part. CS stores a list of strings (messages) in a message list. A message in the message list consists of the following two parts:

- External input
- Classifier execution

The message list is the same element that is referred to as the workspace or blackboard in a production system. There are various possible actions that can be written in the classifier's consequent part, such as updating the message list or reacting to the external environment. In the following description, however, the string in the consequent part is assumed to be placed in the message list as it is. An example of the action of the execution part is as follows (see Fig. 5.22):

**Step1** Message input comes from outside (e.g., 1111).

**Step2** Message list 1011 matches classifier 10##:0100, where # is a wild card symbol that matches both 1 and 0.

**Step3** The consequent part 0100 is added to the message list.

**Step4** The above process is repeated.

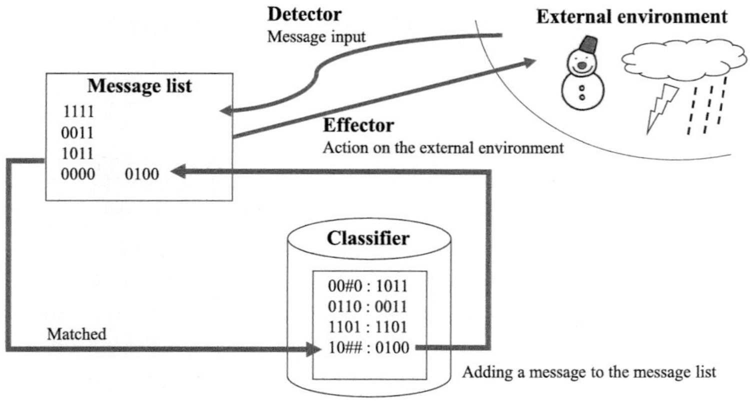


**Figure 5.22:** Example action of classifier system.

The part that acts externally by referring to the message list is called "effector," and the part that inputs messages from the environment is called "detector." Furthermore, a classifier that matches a message is called an "active" classifier. The execution part executes some of these active classifiers.

The key characteristics of two approaches for CS are described as follows:

(1) Michigan approach

Each rule, i.e., classifier, corresponds to one individual in the Michigan approach. The system needs a strengthening functionality that provides a "strength" parameter to each classifier in addition to functionality to execute learned classifiers. Here, strength is a measure of the reliability of classifiers in CSs. Moreover, a functionality to generate new classifiers is necessary, and a GA is used in this generation. The following is an explanation of the functionalities.

- Execution functionality: Searches for a classifier that corresponds to input data (state) from the environment and outputs an action resulting from this classifier to the environment. The classifier to be selected is determined from the strength obtained using past usefulness.
- Strengthening functionality: Observes changes in the environment caused by classifier execution and updates the strength of the classifier. If the result is good, the classifier is determined to be effective and the strength of the executed classifier increases. In contrast, the strength decreases if the result is bad. The system is strengthened to be centered on good classifiers by repeating this process. Proposed learning methods to update the strength include the bucket brigade algorithm and the profit sharing plan.
- Generation functionality: The types of classifiers in a system are finite and the influence on the environment is limited. Generation of new clas-

sifiers is necessary to increase the kinds of action to the environment. However, random generation of classifiers would require too much time before a useful classifier was generated. Therefore, genetic operations are carried out using a GA where classifier is the individual and the strength is the fitness function. New individuals generated by the GA replace individuals in the previous population with low strength or high similarity. Classifiers are generated after every few steps to reflect the strengthening of classifiers based on a combination of many actions to the new classifiers.

(2) Pittsburgh approach

The Pittsburgh approach considers one individual as a set of rules that comprise one function, and can consider genetic operations for each rule, which are each regarded as one unit. Consequently, GAs can be implemented more directly and therefore more easily. The following are the characteristics of the Pittsburgh approach.

- Rules are considered as genes, thus crossover means recombination of rules and mutation means conversion into a different rule. As a result, the rule strengthening functionality is unnecessary, and the implemented GA will be natural and easy to handle.
- The gene length (i.e., the number of rules that make up one individual) is variable, not fixed, because many rules comprise one individual. Therefore, some tricks may be necessary for the gene structure in the GA, such as making the apparent length the same for all genes.
- A large number of rules are necessary to avoid premature convergence. As a result, the number of individuals and the number of rules that each individual contains are very large. Furthermore, evaluation of each individual needs to be carried out for the rule set rather than a simple sum of evaluated values of each rule comprising the individual. Consequently, learning takes time.

Reports of methods that improve classifier systems include the zeroth level CS system (ZCS) [164], the CS system based on accuracy (XCS) [165], and the anticipatory CS system [131].

### 5.4.2 ROIL: Rule Optimization via Large Language Model for Imitation Learning

We introduce a novel approach for solving text-based interactive environments, focusing on transferability, safety, and interpretability. We extend LCS for natural language capability leveraging LLMs. This extension involves human-readable policies, enhancing both understanding and transparency. The method consists of two phases: rule optimization through human demonstrations and action inference for applying learned policies. The extension to LCS and these two phases, together with additional implementation details, are described in the following sections.

**Interactive Tasks** ROIL models the interactive task as a Markov decision process (MDP), following Reinforcement Learning formulation [133]. An *agent* interacts with an *environment* to maximize reward signal. At each timestep $t \in \mathbb{N}$, the agent observes a state $s_t \in S$ and performs an action $a_t \in A_{s_t} \subset A$. The environment responds with a reward $r_t = r(s_t, a_t, s_{t+1}) \in \mathbb{R}$ and a new state $s_{t+1}$, with some state transition probability. The agent's goal is to maximize the cumulative reward $R_t = \sum_{t'=t}^{T} r_{t'}$ over the *episode*. This MDP is defined by the tuple $(S, S_0, A, \rho, r)$, with $S_0$ indicating initial states. Unlike typical tasks with numerical states and actions representation, we focus on text-based interactive tasks, where states and actions are represented in natural language.

**Large Language Models** LLMs represent a significant evolution in natural language processing, characterized by their advanced capabilities in instruction following and context-aware text generation [18, 1]. Finetuning LLMs for specific tasks can be tricky. As opposed to gradient-based methods, we focus on a gradient-free prompt finetuning approach. This approach does not require the developer to have access to the weight of the model, thus, it is more practical for general users, especially when pairing with APIs.

### 5.4.3 Natural Language Learning Classifier System

We have established the Natural Language-based Learning Classifier System (NL-LCS), a modified LCS framework designed for natural language processing in a language-based environment. Our approach distills the LCS to its fundamental components: rule representation, rule optimization, the classifier or matching system, and the action selection mechanism, inspired by the Zeroth level Classifier System (ZCS) [164]. We omit reinforcement feedback to avoid trial-and-error interactions, relying solely on human demonstrations for rule optimization.

The NL-LCS features a policy represented as a human-readable list of rules, which demands natural language interfaces between each module in the system. The rule optimization process will be discussed in the following section. The classifier is replaced with a text classifier, while action selection must also accomodate the natural language observation and the matched rules. These changes, integral to NL-LCS, leverage LLMs for effective functioning in such environment, maintaining our system's simplicity, interpretability, and alignment with safety and transferability goals.

### 5.4.4 Rule optimization

Consider a scenario where we have an observation classifier, denoted as $C$, with classes labeled as $c_1, c_2, ..., c_n$. We can then construct a system for associating a set of rules to each observation class. This set of rules for a class $c$ is defined as the *rule group* $R^c = \{r_1^c, r_2^c, ...\}$. Each rule $r$ is represented as a sequence of tokens. A set of rule groups for a particular environment form a *rule set* $R$, where $R = \{R^{c_1}, R^{c_2}, ..., R^{c_n}\}$ encompasses all rule groups. Rule set can be seen as the agent's "policy". With this formulation in place, it becomes clear that for any given observation categorized as class $c$, there are $|R^c|$ rules within the corresponding rule group that match this observation. This aspect simplifies the matching operation in NL-LCS.

To optimize a rule set, we adopt an iterative improvement approach. Contrary to many previously discussed optimization studies that begin with an initial handcrafted population, our approach starts with an empty rule set. Each iteration involves at most one edit to one rule group within the rule set. Specifically, this editing process in each iteration comprises four main steps: selection of human demonstrations, rule selection, LLM prompting and rule updating as depicted in Fig. 5.23.

**Step 1: Human Demonstrations Selection** The first step of each iteration is the selection of human demonstrations as a source for imitation. The demonstration pool comprised $(state, action)$ tuples retrieved from numerous time steps across various trajectories recorded from human actions. Given that these state-action tuples are static, we can precompute the class of each pair using the observation

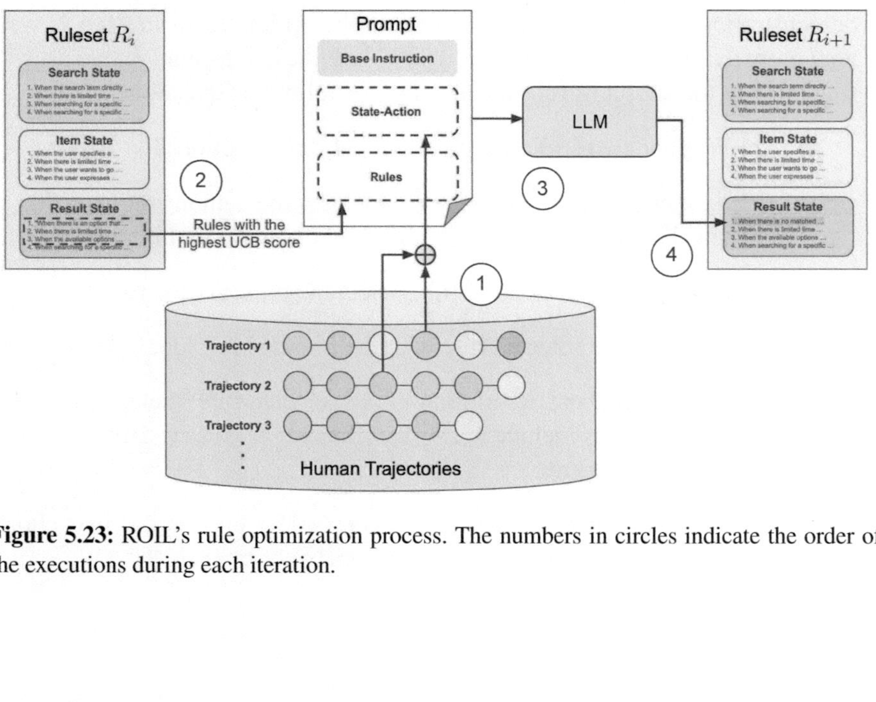


**Figure 5.23:** ROIL's rule optimization process. The numbers in circles indicate the order of the executions during each iteration.

classifier $C$ and record the classifications. We denote these tuples of state-action-class as *demonstrations* $(s, a, c)$.

In this step, we uniformly and randomly selecting $n_d$ demonstrations with the same classes but different actions from the pool. Our preliminary experiments indicate that relying on a single demonstration for rule inference leads to task-specific rules lacking transferability, while incorporating multiple demonstrations, despite higher computational costs, enhances the generality of the rule set. The demonstration selection strategy stems from the idea of adversarial examples [43], the underlying rule must be expressive enough to distinguish similar observations leading to each unique action. In our experiments, we consistently use $n_d = 2$, which is constrained by the maximum context length of the LLM employed.

**Step 2: Rule Selection** Considering the limited context length constraint, we cannot optimize every rule in a rule group at once. It is important to balance between efficiently exploring the rule space and exploiting the existing rules. We utilize the Upper Confidence Bound (UCB) algorithm to select a subset of the rule group to be optimized. We maintain records of each rule's applicability and the selection frequency denoted as $A(r_i^c) \in \mathbb{N}$ and $N(r_i^c) \in \mathbb{N}$, where $i$ is the iteration number. The UCB score is computed using eq,(5.1). The top 10 rules with the highest UCB score are selected for optimization.

$$U(r_i^c) = \frac{A(r_i^c)}{N(r_i^c)} + \sqrt{\frac{2\log i}{N(r_i^c)}} \tag{5.1}$$

**Step 3: LLM Prompting** Having gathered all necessary ingredients, including demonstrations from **Step 1** and the rules to be optimized from **Step 2**, we are now prepared for the prompting step. Providing these information, we asked the LLM to edit the subset of rule group using the following questions:

- **Q1** Why the two scenarios resulted in different actions?
- **Q2** Extract one rule behind the reason why the demonstration action was chosen. The rule must be generalized, not specific to this example.
- **Q3** From the list of existing rule, select the most related rule.
- **Q4** Compare the selected rule with the founded key idea/rule.
- **Q5** What is the best description of the selected rule? Select one of the 3 choices? Choices include `same_intention`, `similar_intention`, and `different_intention`.
- **Q5.1** If you chose `same_intention`, select one of the two choices: `found_rule_better` or `selected_rule_better`.

**Step 4: Rule Updating** Suppose that at the $i$th iteration, we select demonstrations belonging to class $c$ and rules $r^c_{i,1}, r^c_{i,2}, \ldots, r^c_{i,n_r}$ from the rule group $R^c_i$ as determined in **Step 1** and **Step 2**. After **Step3**, we acquire a new candidate rule $r'$ from Q2, the most relevant rule to the demonstration $r^c_{i,j}$, with $j \in \{1,2,\ldots,n_r\}$, from Q3. Finally, the comparison result is given from Q5, which is one of four scenarios: `found_rule_better`, `selected_rule_better`, `similar_intention`, or `different_intention`. The outcome of Q5 determines the edit operation of the current iteration rule set as follows:

$$\begin{aligned}
&\texttt{found_rule_better}: & r^c_{i+1,j} &\leftarrow r' \\
&\texttt{selected_rule_better}: & r^c_{i+1,j} &\leftarrow r^c_{i,j} \\
&\texttt{similar_intention}: & r^c_{i+1,j} &\leftarrow r' \\
&\texttt{different_intention}: & R^c_{i+1} &\leftarrow R^c_{i+1} \cup \{r'\}
\end{aligned}$$

The newly added rule in `different_intention` case always have applicability and selection frequency initiated to 1. The applicability of the updated rule for other cases is always incremented by 1, i.e., $A(r^c_{i+1,j}) = A(r^c_{i,j}) + 1$. This is because in all cases, we only update the rule if the generated one is better in the comparison. The selection frequency for other cases is updated according to eqs. (5.2) and (5.3).

$$N(r^c_{i+1,k}) = N(r^c_{i,k}) + 1 \quad \forall k \in \{1,2,\ldots,n_r\} \setminus \{j\} \tag{5.2}$$

$$N(r^c_{i+1,j}) = \begin{cases} 1 & \texttt{found\_rule\_better} \\ N(r^c_{i,j}) + 1 & \text{otherwise} \end{cases} \tag{5.3}$$

## *5.4.5 Action inference*

After successfully optimizing the rule set for our NL-LCS, the action inference phase becomes quite straightforward as shown in Fig. 5.24. Upon recieving a new observation, it is being classified as a class, $c$. Subsequently, the top 10 rules from the rule group $R^c$ within the rule set $R$ are selected based on their applicability, $A$. This selection of the top rules can be viewed as a coarse-matching process between the observations and the rules. In this phase, we utilize the LLM to enhance this matching process implicitly. This is achieved by inducing a thought process response, akin to the approaches of some previous studies [72, 158]. The generated thought process is then being integrated into the action selection step. In this phase we use the following questions:

- **Q1** Based on the observation, what's your thought on selecting an action? Use applicable rules from the provided rule list as a source for reasoning. Think step by step.
- **Q2** Given the selected rule, select the most relevant action.

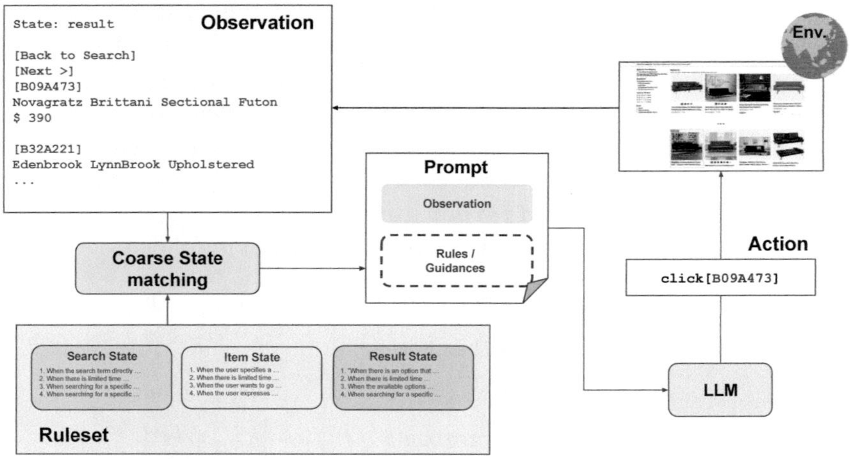


**Figure 5.24:** Action inference loop of ROIL.

Additionally, there are finer details necessary for practical implementation including state augmentation and error handling. Firstly, in addition to the selected rules and the current observation, we supply three important informations: previously taken actions, available actions, and time limitation signal. The list of previously taken actions and available actions guides the LLMs in selecting the most relevant action from the allowed options at hand. The time limitation signal is simply a string, `Timestep left: low`, appended to the beginning of the observation when the number of time steps remaining reaches the threshold of 2. This is important for the evaluation, as the agent learned from human demonstrations may adopt a lengthy decision–making process, exceeding the time limit and resulting in an error in performance measurement.

Secondly, running an automated system that involves LLMs' API calls can be prone to instability. Two common errors we encounter are formatting errors and API failures. To combat both type of errors, we utilize parallel asynchronous API calls during the LLMs utilization for both rule optimization and action inference phases. Additionally, we have set a maximum retry limit of 2 for each API call in the event of failure.

## 5.4.6 Experimental set-ups

The experiments are designed to answer three research questions:

- How well does ROIL perform compared to related methods?
- Can LLM be used as an optimizer for the agent's policy?
- And finally, do the ROIL agents behave similarly to the humans they learned from?

**Table 5.6:** Hyperparameter configurations for the evaluation experiments.

| Hyperparameter Name | Value |
|---|---|
| low time step threshold | 2 |
| max episode length | 15 |
| rule group subset size | 10 |
| webshop instruction numbers | 0-49 |
| API timeout (seconds) | 50 |
| API max retry | 2 |
| LLM temperature (optimization) | 0.6 |
| LLM temperature (evaluation) | 0 |

To address these questions, we adopt WebShop environment [170]. WebShop is a text-based environment simulating an e-commerce web interface with 1.1 million products. It contains 4 different pages including *search*, *results*, *item*, and *item_description*. In each episode, a random instruction to find a specific item is given. The goal is to navigate through the website to find and purchase a product that matches with the given instructed description. Each instruction can be seen as a unique task under the same environment. The action space of the environment ranged from button clicking to arbitrary text input. Alternatively, text-based environment benchmarks such as [121, 173, 142] can also be used. However, among these environments, WebShop presents the most challenges according to prior works [120, 169, 176] and also readily provides human demonstrations.

WebShop comes with 12,087 predefined instructions, distributed into 10,587/1,000/500 entries of train/development/test splits. Together with the environment is the 1,012 trajectories of human demonstration collected from participants who were asked to complete the task using a subset of instruction from train split. We further filtered the collected demonstration so that only the successful trajectories are used during the rule set optimization step. The impact of the demonstration quality will be discussed in the following section. WebShop benchmark has two metrics, i.e., *score* and *success rate*. The *score* is the percentage of matched attributes between the purchased item and the instructed description with the maximum of 100 per task. A task is determined to be *success* if the purchased item is perfectly matched the instruction, with the score of 100.

Due to the resource limitation, all experiments were conducted with three seeds, using the first 10% or 50 instructions of the test split. We used `chat-gpt-turbo-0613`,[2] a conversational text completion large language model, as the LLM for both rule optimization and action inference. We set the hyperparameters as shown in Table 5.6. To maintain simplicity without intro-

[2]OpenAI's API: platform.openai.com/docs/guides/text-generation/chat-completions-api (Accessed October 1st, 2023).

ducing another moving part, we utilize the environment's predefined observation classification. This results in four types of observation classes, each of which corresponds to one web page. Consequently, there will be four rule groups within the rule set.

### 5.4.7 Performance comparison

**Baseline Comparison** ReAct [169] is used as a strong baseline due to its similar requirements for evaluation. Since the LLM utilized in ReAct differs from ours, we re-evaluated ReAct under the same settings and test split. The re-evaluation was conducted with minimal modification to the method to ensure the consistency and comparability when a different LLM is used. Reflexion [120] also qualifies as a baseline. However, due to their reports of the comparable performance with ReAct on WebShop environment, we decided to omit the re-evaluation of Reflexion. We also report the performance of the human expert and traditional imitation learning (IL) taken from [170] as a reference. Note that these two references were evaluated with full test split of 500 instructions.

The main results are shown in Fig. 5.25. The main ROIL agent, labeled with *ours*, has performance improvement over ReAct by around 50% for both metrics. This indicates the quality of the optimized rules while showing that the formulation of prompt optimization, as rule set and rule group, of ROIL for NL-LCS has positive impact on the action inference with LLM. Since the evaluation is done on unseen tasks, the positive performance also indicate the transferability property of ROIL. With that in mind, ROIL aims to imitate human decision making, yet it fails to reach human expert level of performance, although it shows a very close performance to gradient-based imitation learning.

**Ablation Study** We further demonstrate the effectiveness of ROIL through two ablation studies for a conclusive analysis. First, we conducted a performance evaluation of ROIL compared to an LLM agent. This agent, labeled as *w/o rules* in Fig. 5.25, utilized the same action inference prompting and observation augmentation method, but without the incorporation of rules. The primary difference between this agent and the main agent is the presence of rules. The results indicate that the learned rules are the main factor to performance enhancement, thereby validating that the effectiveness of ROIL is attributed to the optimization process, and not the carefully crafted action inference prompt. Second, we assessed the performance of ROIL after optimization from unfiltered demonstrations. The human demonstrations sourced from [170] were unpolished, each trajectory containing random explorations. For our main agent, these demonstrations were refined by omitting parts that did not contribute to progress. We compared this main agent against an agent optimized using the unfiltered demonstrations, denoted as *unfiltered data* in Fig. 5.25. The stark disparity in the performance between the two agents highlight the significance of demonstration

**Table 5.7:** The average amounts of tokens used for rule set optimization and inference with ReAct and ROIL.

| Method | Tokens used | |
|---|---|---|
| | Optimization (per iteration) | Inference (per timestep) |
| ReAct | N/A | $1376 \pm 18$ |
| ROIL | $2317 \pm 63$ | $953 \pm 29$ |

quality. Additionally, this finding further confirms that rule optimization is the primary contributor to the performance gain.

**Efficiency Comparison** In addition to the score and success rate metrics, we measured the amount of tokens used by ROIL compared to ReAct. The results are given in Table 5.7. Although ReAct does not have optimization step, it requires a full successful trajectory as a static example in the prompt. Furthermore, ReAct also includes recent conversations in its prompt for additional context. These two factors resulted in a much higher token usage during inference. In contrast, ROIL only requires a single rule set as a prompt in addition to the base instruction. This resulted in 30% reduction in token usage during inference compared to ReAct. The efficiency in token usage opens up the possibility of scaling up to a more complex environment and a more complex prompt formulation. We have limitted the computational efficiency comparison to ReAct due to the comparability of the two. Although ROIL requires significantly less data for the optimization process to achieve the result in Fig. 5.25 compared to the training process of a traditional imitation learning agent, a more comprehensive study is needed for a quantitative data efficiency comparison.

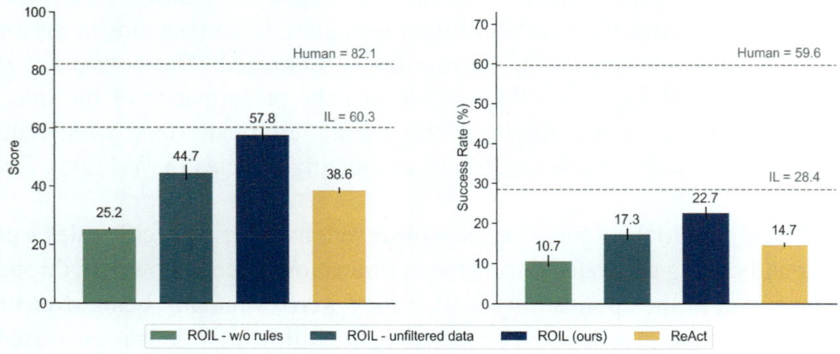


**Figure 5.25:** Average score (left) and success rate (right) of ROIL compared to LLM evaluated on the same inference prompt without the rules, ROIL optimized using unfiltered demonstration, and ReAct.

### 5.4.8 Behavior analysis

To further understand the behavior of ROIL, we analyzed the frequency of each state visitation. The average number of some of the states visitation frequency per episode are presented together with the averaged total state count per episode in Table 5.8. We can see that ROIL spends almost the same amount of time in search state which is more than 1.0, hinting a *re-querying* strategy resembling the human behavior. However, a big disparity in the item state count and the total state count can be observed, suggesting that ROIL is not perfectly imitating the human behavior. It is worth noting that the reference numbers are from full test split, and we also enforced time step limitation of 15 during the evaluation of our method which could partially be the cause of the differences.

**Table 5.8:** The average number of state visitation per episode for total state count, search state, and item state of WebShop environment.

| Method | State Count | | |
|---|---|---|---|
| | Total | Search | Item |
| Human | 11.3 | 1.4 | 1.9 |
| IL | 9.4 | 1.6 | 1.3 |
| ROIL | $7.2 \pm 0.9$ | $1.4 \pm 0.1$ | $2.5 \pm 0.4$ |

### 5.4.9 Using Large Language Model as an optimizer

Contrast to the gradient-based optimization methods, there are relatively small numbers of studies that specifically analyze the use of LLM as an optimizer for optimizing policies. We conducted a simple experiment to understand the improvement generated by a LLM during the optimization process. Specifically, we evaluate the performance of the rule set at various optimization iterations. It is worth noting that the number of iterations directly corresponds to the amount of demonstrations from which a rule set is optimized. The results are shown in Fig. 5.26 and Fig. 5.27. We can see that the performance of the rule set is gradually improved over the iterations proving emirically that, with ROIL, the employed ChatGPT-3.5 is capable of optimizing policies in WebShop environment.

In addition to the figures, we also observed that overly specific rules were occasionally being generated during the optimization process, however, through the comparison of the applicability values, they were eventually being overruled as the optimization progressed. For instance, the following rule was generated during the optimization process for the search state: "When the instruction specifies the search criteria for a gold floor lamp with a glass shade and a nickel finish, the best action to take is to search for a product that fulfills these criteria." This rule has an applicability value of 1, indicating that it was immediately disregarded in favor of a more generalized rule in the subsequent iterations.

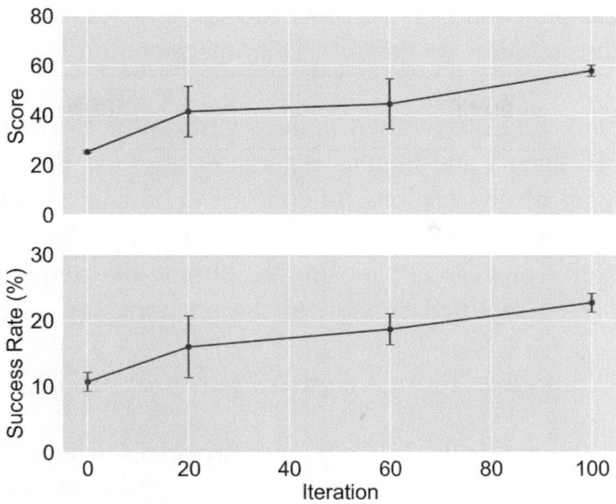


**Figure 5.26:** The score and success rate of ROIL at different optimization iterations.

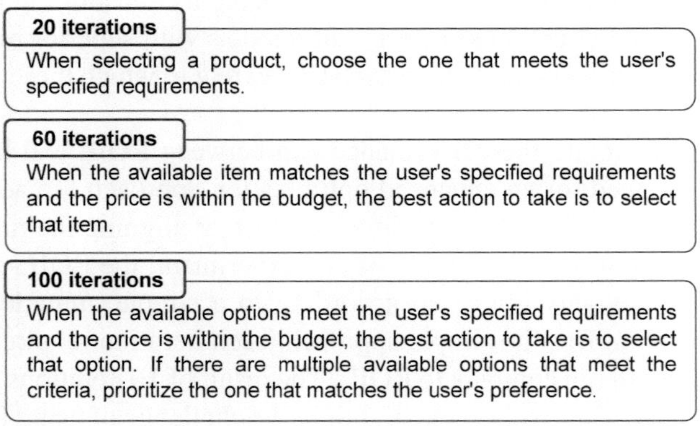


**Figure 5.27:** An example of rule evolution hand-picked from *result* rule group over different iterations.

## 5.4.10 Discussion

One intriguing question is, what is the primary factor to the quality of the rules learned by ROIL. We hypothesized that ROIL is able to learn or 'discover' high-quality rules if quality demonstrations are provided. This is vaguely confirmed in the second ablation study, where the performance of ROIL, when optimized with unfiltered demonstration, is significantly lower compared to that optimized from filtered demonstration. In other words, the one with higher chance of encountering quality demonstrations has a better performance. One more subtle evidence is in the gradual variance reduction of the performance in Fig. 5.26. One interpretation is that there are prime demonstrations that give a quality insights for the rule set, and the chance of encountering those demonstrations increases as the optimization iteration progresses. This interpretation is inline with the hypothesis.

Another interesting observation is the significant performance disparity between our ROIL agents and human experts, despite the ROIL agents learning rules from human demonstrations. In contrast to human cognition, these rules, generated through the LLM's inductive reasoning from human actions, fail to emulate the reverse process of choosing the human-like actions. For clarity, we can formulate the simplified version of the phenomenon as an inequation in eq. (5.4):

$$\underset{a}{\text{argmax}}\, P_{act}(a|s, \underset{r}{\text{argmax}}\, P_{gen}(r|s,a^*)) \neq a^*, \tag{5.4}$$

where $P_{gen}$ and $P_{act}$ are the LLM's inference processes using rule generation and action selection template, respectively. $a^*$ is the groundtruth action, $s$ is the state observation, and $r$ is the rule. This shortfall might be attributed to the model's incapability of coherent inductive-deductive reasoning. Thus, we term this phenomenon *the incoherency of the reverse generation* of LLMs. To investigate this problem further, experiments could be conducted by substituting the LLM used in rule generation or action selection with models with superior reasoning capabilities. Such studies would help to assert the extent to which these factors impact the performance gap.

With that in mind, the performance gap between ROIL and human experts highlights the need for an enhanced optimization approach. To address this, we have conducted an extended experiment on further aligning the rule set with the human demonstration. Specifically in this experiment, the ROIL-optimized rule set is being iteratively improved according to a simple, mutation-only evolutionary algorithm. Despite simple approach, each iteration requires a magnitude more resources to run compare to ROIL. The result of 10 iterations shows no sign of improvement. This experiment highlight the challenge of using LLM as an optimizer in this area, and a carefully designed approach is necessary to achieve a human-level performance.

We have proposed a method for integrating LLMs with LCS for interactive tasks, leaving out several areas that can be explored further. These include:

- **Demonstration Selection Strategy:** Our current method employs a basic random selection of adversarial examples. Future work could explore more sophisticated strategies, such as state-dependent sampling, to enhance the effectiveness of the learning process.

- **Optimization Scheme:** We have utilized a simple iterative improvement technique which is still far from human-level performance Alternative approaches like meta-heuristic based algorithms present promising alternatives for further investigation.

- **Other Considerations:** Additional factors, such as the choice of the LLM, the choice of environment for the interactive task, and the choice of observation classifier used, also offer opportunities for refinement and experimentation in future studies.

These aspects present potential enhancements to ROIL that could be further investigated.

# Chapter 6

# Emergence of Intelligence for AI

> Familiarity breeds liking.
>
> People who face very bad options take desperate gambles, accepting a high probability of making things worse in exchange for a small hope of avoiding a large loss. Risk-taking of this kind often turns manageable failures into disasters. (*Thinking, Fast and Slow* by Daniel Kahneman)

## 6.1 Decoy advertisement

Let us divide one group randomly into two sub-groups and ask each a question A or B.

**Question A**
You are about to subscribe to an annual subscription to your favorite magazine or newspaper. Suppose you have the following choices. Which of these would you prefer? Make your decision as intuitively as possible.

**Option A-1** One year free access to the web version, WWW only subscription $6 (per month)

**Option A-2** Print-only subscription for one year $12 (per month)

**Option A-3** One year subscription to print and web edition of WWW $12 (per month)

**Question B**

You are about to purchase an annual subscription to your favorite magazine or newspaper. Suppose you have the following choices. Which of these would you prefer? Make your decision as intuitively as possible.

**Option B-1** One year free access to the web version, WWW only subscription $6 (per month)

**Option B-2** One year subscription to print and web edition of WWW $12 (per month)

The publisher (the provider of the magazine or newspaper) wants to make as much profit as possible, so it wants the customer to choose Option B-2, which is identical to Option A-3. On the other hand, Option A-2 is worthless because the content is inferior to Option A-3 for the same amount of money. Therefore, if we make rational decisions, the percentage of choosing Option B-2 (=Option A-3) will be about the same for both Questions A and B.

In fact, the results of a survey conducted with the author's college students are shown in Table 6.1.[1] As can be seen from the table, the percentages of choosing Option A-3 and Option B-2 for questions A and B are very different. Why is this?

**Table 6.1:** Percentage of responses to questions A and B.

| | A-1 (%) | A-2 (%) | A-3 (%) | Total no of Q.A | B-1 (%) | B-2 (%) | Total no. of Q.B | Total no. of group |
|---|---|---|---|---|---|---|---|---|
| group1 | 58.3 | 4.2 | 37.5 | 48 | 82.4 | 17.6 | 51 | 99 |
| group2 | 61.9 | 2.4 | 35.7 | 42 | 72.7 | 27.3 | 44 | 86 |
| group3 | 65.7 | 3.0 | 31.3 | 67 | 80.0 | 20.0 | 55 | 122 |
| group4 | 78.8 | 3.9 | 17.3 | 52 | 88.9 | 11.1 | 54 | 106 |

Question A is a decoy advertisement. It is known that a consumer who is unsure which of two options to choose is led to one of the original options by being presented with a third option (i.e., the discarded option). In the two questions above, question A contains a decoy advertisement (Option A-2). The content of questions A and B are identical except for the presence of the decoy. The results in Table 6.1 show that the selection rate for Option A-2 is clearly higher than that for Option B-2 in Question B without the decoy advertisement.

[1]The subjects of the experiments described in this chapter are primarily undergraduate and graduate students at the author's university (ages 22 to 25, approximately 70% male to female).

## 6.2 Bernoulli's misunderstanding

In Section 4.4, we explained the problems with classical utility theory. In fact, utility does not make sense to many people because risk cannot be captured by a single number (i.e., real-valued expected utility). It is unclear when the utility function should have a logarithm or square root. It is also believed that humans do not have a natural utility function.

To begin with, let us consider how humans do not follow the utility axiom in page 114. For example, assume the following situation:

**Case1**: You receive \$1,000 in addition to your current wealth and are asked to choose between the following two:

- (a) 50% chance of getting \$1,000, but 50% chance of getting nothing, or
- (b) you will receive \$500 with certainty.

Which would you choose?

In this case, experiments have shown that most people will choose option (b), i.e., you will surely receive \$500. Now forget this question and consider the following situation:

**Case2**: You receive \$2,000 in addition to your current wealth and are asked to choose between the following two:

- (a) 50% chance of losing \$1,000, but 50% chance of losing nothing, or
- (b) you will loose \$500 with certainty.

Which would you choose?

In this case, most people would choose to lose \$1,000 with a 50% probability, i.e., option (a).

Now compare these two situations. In both cases, the probability for option (a) is the same, i.e., \$1,000 or \$2,000. The price will certainly be \$1,500 for option (b). According to Bernoulli's expected utility theory, options (a) and (b) are exactly the same in both cases. Therefore, a person who chose (a) in **Case1** should choose (a) in **Case2**, and a person who chose (b) in **Case1** should choose (b) in **Case2**. However, this was not the case. Many people chose (b) in **Case1**, and many people chose (a) in **Case2**. In other words, many people chose certainty in **Case1**, while many people chose gambling in **Case2**. Why did this happen? Where did Bernoulli go wrong?

Bernoulli defined utility by assuming that the gain from any small increase in wealth should be inversely proportional to the amount of wealth held in the past. For example, the logarithmic utility function described in Fig. 4.1. However, this is not necessarily true. It seems that the degree of happiness is determined by the change in the initial state of wealth. The initial wealth is called "reference point." In other words, the evaluation of an adventurous opportunity (bet) depends more on the reference point at which the possible gain or loss occurs than on the resulting wealth.

**Definition 6.1 Reference point**

- Reference point is the initial state of wealth.
- Degree of happiness is determined by the change from the reference point.
- Preference is manipulated by the change of the reference point.

Consider the two cases just described. In **Case1**, the reference point is +$1,000, since you have received $1,000 in addition to your current wealth. Therefore, an increase of $500 is a good thing. In **Case2**, on the other hand, the reference point is +$2,000 so that a definite decrease of $500 is not desired. As we will see later, this conclusion follows from the fact that humans have different attitudes toward losses and gains (i.e., risk-aversive and risk-seeking in Fig. 4.6).

An example of the errors of classical utility theory is the inconsistency of attitudes toward betting. Consider the following two bets:

**Bet1** 50% chance of losing $100, while 50% chance of getting $200.

**Bet2** 50% chance of losing $200, while 50% chance of getting $20,000.

Many people may not take **Bet1**. However, classical utility theory would lead us to believe that a person who refuses to take **Bet1** will also refuse to take **Bet2**. But this is not possible. No sane person would ever refuse **Bet2**. This can be shown as follows.

Consider a person who avoids a bet that has a 50% chance of losing $100 but a 50% chance of getting $200. Let $U(x)$ denote this person's utility function for $x$ dollars. At this point, we may assume the following two utility functions:

- monotonically increasing, i.e., $U'(x) > 0$,
- convex up, i.e., $U''(x) < 0$.

The second property means that the utility function has diminishing marginal utility (see page 115). Assuming that the previous person has $w$ dollars, the following follows from the fact that he avoids betting:

$$U(w) > 0.5 \times U(w-100) + 0.5 \times U(w+200)$$

This is because the current utility of his wealth is greater than the expected value of the utility after the bet is made. Since $U$ is convex upward, the slope of $U(x)$ decreases as $x$ increases, leading to the following inequality:

$$U(w+100) > 0.5 \times U(w) + 0.5 \times U(w+200).$$

Adding these two equations together yields:

$$\frac{1}{2}(U(w) - U(w-100)) > U(w+200) - U(w+100).$$

Since this holds for any $w$, setting $w$ to $w-100$ yields:

$$\frac{1}{2}(U(w-100) - U(w-200)) > U(w+100) - U(w).$$

Adding these two equations together yields the following equation:

$$\frac{1}{2}(U(w) - U(w-200)) > U(w+200) - U(w).$$

Repeating this equation, we obtain the following equation:

$$\begin{aligned}
(\frac{1}{2})^2(U(w) - U(w-200)) &> U(w+200\times 2) - U(w+200), \\
(\frac{1}{2})^3(U(w) - U(w-200)) &> U(w+200\times 3) - U(w+200\times 2), \\
&\cdots \\
(\frac{1}{2})^k(U(w) - U(w-200)) &> U(w+200\times k) - U(w+200\times (k-1)),
\end{aligned}$$

which shows that the additional utility due to an increase of \$200 is decreasing exponentially. Then, evaluating the incremental value of receiving \$20,000, we obtain the following equation:

$$\begin{aligned}
U(w+20000) - U(w) &= (U(w+20000) - U(w+199800)), \\
&\quad +(U(w+199800) - U(w+199600)), \\
&\quad +\cdots, \\
&\quad +(U(w+200) - U(w)), \\
&< (\frac{1}{2} + \frac{1}{2^2} + \cdots \frac{1}{2^{100}})(U(w) - U(w-200)), \\
&= (1 - \frac{1}{2^{100}})(U(w) - U(w-200)), \\
&< U(w) - U(w-200).
\end{aligned}$$

**Table 6.2:** Maximum amount to turn down a bet that loses $L$ at 50% for those who turn down a bet that $g$ at 50% and loses $100 at 50%.

| | $g$ | | | |
|---|---|---|---|---|
| $L$ | 101 | 105 | 110 | 125 |
| 400 | 400 | 420 | 550 | 1,250 |
| 600 | 600 | 730 | 990 | ∞ |
| 800 | 800 | 1,050 | 2,090 | ∞ |
| 1,000 | 1,010 | 1,570 | ∞ | ∞ |
| 2,000 | 2,320 | ∞ | ∞ | ∞ |
| 4,000 | 5,750 | ∞ | ∞ | ∞ |
| 6,000 | 11,810 | ∞ | ∞ | ∞ |
| 8,000 | 34,940 | ∞ | ∞ | ∞ |
| 10,000 | ∞ | ∞ | ∞ | ∞ |
| 20,000 | ∞ | ∞ | ∞ | ∞ |

From this, we have the following equation:

$$U(w) > 0.5 \times U(w-200) + 0.5 \times U(w+20000).$$

Thus, it can be shown that we do not participate in **Bet2**.

The crux of the proof is that it interprets a bet of about $100 in terms of a utility function over a larger range. As will be explained below, to approximate human thinking, one must take into account that it is more sensitive to losses than gains.

By making the same derivation as above, we obtain results such as those in Table 6.2. The table shows the maximum amount by which a person who turns down a bet that gains $g at 50% and loses $100 at 50% will turn down a bet that loses $L$ at 50%. For example, a person who turns down a bet that he gains $110 in 50% and loses $100 in 50% will also turn down a bet that he gains $2,090 in 50% and loses $800 in 50%. He would also turn down a bet that he would lose $1,000 in 50%, no matter how big he could get in 50%.

Now let me explain the paradox of the French economist Maurice Allais (see Fig. 6.1). First consider which of the following jar you would choose:

**Jar A** includes 100 $1M winning lottery tickets.

**Jar B** includes:

- 10 $5M winning lottery tickets,
- 89 $1M winning lottery tickets,
- 1 blank ticket.

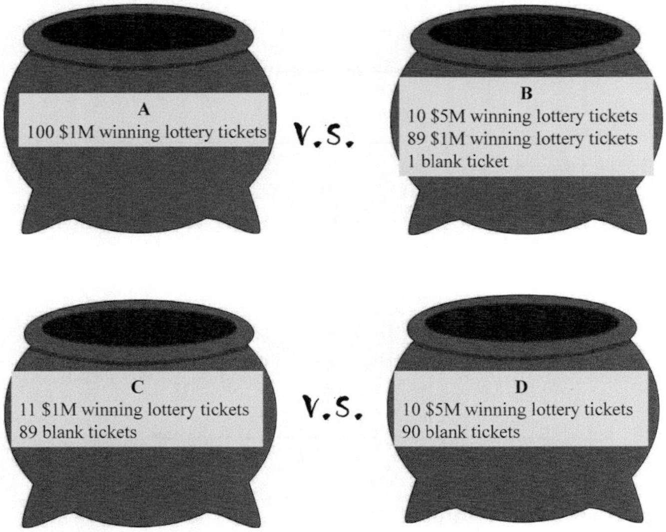


**Figure 6.1:** Which jar would you choose?

Next, forget about this question and answer which jar you would choose:

**Jar C** includes:

- 11 $1M winning lottery tickets,
- 89 blank tickets.

**Jar D** includes:

- 10 $5M winning lottery tickets,
- 90 blank tickets.

According to expected utility theory, it must be the case that:

- A person who chooses **Jar A** chooses **Jar C**,
- A person who chooses **Jar B** chooses **Jar D**.

The proof is as follows. Assume that $U(\$5\text{M}) = 1$, $U(\$1\text{M}) = a$ and $U(0) = 0$. Let $U_X$ be the utility of jar X. Since $U_A = a$, $U_B = 0.1 \times 1 + 0.89a$, $U_C = 0.11a$, $U_D = 0.1$, we have the following results:

$$A \succeq B \iff U_A \geq U_B \iff 0.11a \geq 0.1$$
$$C \succeq D \iff U_C \geq U_D \iff 0.11a \geq 0.1$$

Therefore, it is a contradiction if both $A \succeq B$ and $D \succeq C$ are satisfied.

However, Allais predicted that most humans would be $A \succeq B$ and $D \succeq C$. In Tversky's experiment, 17 of the 29 participants showed the same preference pattern as Allais's prediction. More interestingly, after the classical utility theory was taught, they disagreed again, and all 17 did not change their opinions. Thus, we have a case that does not satisfy the axiom of a rational agent. In orther words, the existence of a completely rational agent has become doubtful.

Therefore, modern economics, which assumes the existence of rational agents based on perfect information and expected utility, was forced to be revised. Classical economics has tended to neglect psychology, even though it is a human activity. In response, the field of "behavioral economics," which actively incorporates behavior psychology, was born. Behavioral economics starts from the assumption that humans are sometimes not rational at all and do not act according to classical utility theory. The impetus for this was the prospect theory of Kahneman[2] and Tversky.[3] This theory assumes that there are no rational agents (called "econ") and attempts to explain human behavior by the following elements [62]:

- loss aversion
- present bias
- endowment effect (divestiture aversion)

These are explained in detail below.

## 6.3 Endowment effect: Why is a money-back guarantee profitable?

Let us divide the students into two groups A and B [61]. The students in Group A are then given university mugs (see Fig. 6.2). After a while, they are asked how much they would sell the mug. On the other hand, the students in Group B do not receive a mug. The students who did not receive the mug are asked how much they would pay to get it. Which of the following cases is realized?

- The student with the mug rated it higher.
- The student without the mug rated it higher.
- Both students rated it equally.

---

[2] Daniel Kahneman (1934–): American psychologist. He proposed behavioral finance theory and prospect theory, which integrate economics and cognitive science. He was awarded the Nobel Prize in Economics in 2002.

[3] Amos Tversky (1937–1996): Israeli-born psychologist. He developed prospect theory in collaboration with Kahneman.

**Figure 6.2:** Endowment effect (1).

Rationally, both students should rate the same thing the same way. However, the average sale price for the student with the mug was \$4.50, while the average payment for the student without the mug was only \$2.25.

In another experiment, we divide the students into three groups A, B, and C (Fig. 6.3) [70].

- The students in group A are first given a mug. After that, they are offered to exchange it for a chocolate bar.
- The students in Group B are given a chocolate bar. They are then offered the choice of exchanging it for a mug.
- The student in Group C is offered either a chocolate bar or a mug from the beginning.

Which group of students is most likely to choose the mug?

The results show that 89% of the students in Group A chose the mug, 10% in Group B, and 59% in Group C. This result can be explained as the "endowment effect." This means that people value more what is theirs.

**Definition 6.2 Endowment effect**
Ownership creates satisfaction.

Interestingly, the endowment effect has also been found experimentally in chimpanzees. In this experiment, 33 chimpanzees were given either a popsicle or peanut butter to exchange for the other (see Fig. 6.4). Both popsicles and peanut butter are the chimpanzees' favorite food. As predicted by the endowment effect, 89% of the chimpanzees chose peanut butter if it came first. On the other hand, 42% of the chimpanzees chose the popsicle.

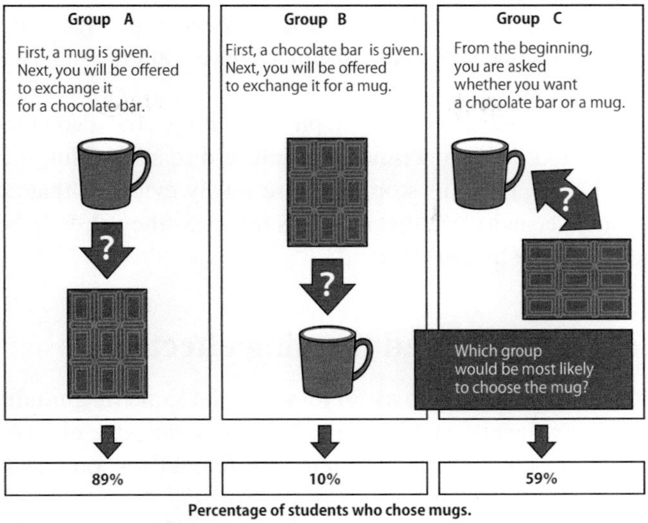


**Figure 6.3:** Endowment effect (2).

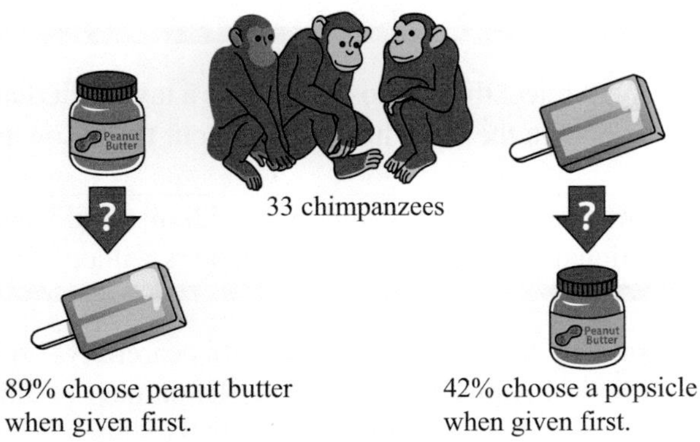


**Figure 6.4:** Endowment effect (3).

The endowment effect is frequently used in marketing. For example, why are money-back guarantees profitable? The money-back effect increases the likelihood of closing a deal once a customer tries a product. It also reduces the hesitation to try the product. A similar efficacy is also known as the IKEA[4] effect [91]. This is based on the cognitive bias that people become attached to things they have made. As people invest more resources, their sense of ownership increases. Marketing also uses a virtual sense of ownership. When people imagine owning or living with a product, they feel as if they already own it. This sense of virtual

[4] A mass furniture retailer that originated in Sweden and has opened stores all over the world. At IKEA, customers often create their own furniture by making small-scale crafts.

ownership increases the likelihood of inducement to purchase or actually own the product. Thus, successful advertisements excel at creating vivid imagery of life with the product.

The giving effect exists not only in physical things, but also in thoughts. Once an idea is accepted, it is sometimes difficult to discard it. This is the beginning of ideological rigor. Humans stop taking seriously evidence that contradicts the idea. They only see evidence that supports the ideas they already believe. As the saying goes, "First come, first served."

## 6.4 Loss aversion and framing effects

Humans are more motivated to avoid losses of the same magnitude than to gain. This is called "loss aversion." In particular, when the same choice is referred to (framed) in terms of loss rather than gain, a different decision may be made. This is called the "framing effect."

For example, consider the following proposal [62]:

> **Proposal1**: Deductions for children should be higher for high-income earners than for low-income earners.

Many people will reject this proposal, because a tax credit that favors the rich is out of the question. On the other hand, what about the following proposal?

> **Proposal2**: Low-income earners without children should pay the same amount of additional taxes as high-income earners without children.

They will also deny it. But let us consider this carefully. Logic dictates that we cannot reject both. **Proposal 2** is based on the number of children, which is the same as taxing households with fewer than two children, for example. Now, if **Proposal 1** requires that low-income individuals should receive the same (or more) deduction for having children as high-income individuals, then **Proposal 2** requires that they pay at least the same amount of additional tax for not having children.

As another example of the framing effect, consider the following question [146]:

> Suppose 600 people are exposed to a deadly infectious disease. Which of the following two options would you choose?
>
> **Option 1** 400 people will die.
>
> **Option 2** 1/3 chance 600 people will be saved, 2/3 chance none will be saved.

In this case, most people choose the first one (72% for **Option 1** vs. 28% for **Option 2**).

Now consider the following question:

> Suppose 600 people are exposed to a deadly infectious disease. Which of the following two options would you choose?
>
> **Option 3** 200 people will be saved.
>
> **Option 4** 1/3 chance no one dies, 2/3 chance 600 people die.

In this case, most people choose the second one (22% for **Option 3** vs. 78% **Option 4**).

This result raises the question of human rationality. The following two cases have the same utility:

- 1/3 chance no one dies, 2/3 chance 600 people die.
- 1/3 chance 600 people will be saved, 2/3 chance none will be saved.

Humans, however, chose the former 78% of the time, and only 28% the latter. This indicates that humans take a great deal of risk in avoiding losses. In other words, describing the same choice as a loss changes the decision.

As an example of loss aversion, consider the following question:

> Which of the following two options would you choose?
>
> **Option 1** You get $240 unconditionally.
>
> **Option 2** 25% chance of getting $1,000, 75% chance of getting nothing.

In this case, the first choice was 84%. This means that people are unlikely to take a risk for an extra gain.

Now consider the following question:

> Which of the following two options would you choose?
>
> **Option 1** You lose $750 unconditionally.
>
> **Option 2** 25% probability of losing $1,000, 75% probability of losing nothing.

In this case, 87% choose the latter. In other words, they are more likely to take a risk to avoid an extra loss.

## 6.5 Utility functions explaining human cognition

The effects we have seen so far, i.e., loss aversion and reference points, cannot be interpreted with classical utility functions.

Therefore, Kahneman and Tversky devised a utility function that is asymmetric and has a steeper slope for losses than for gains. This is called the Kahneman-Tversky value function (see Fig. 6.5). This function is based on the fact that people treat gains and losses asymmetrically and, in particular, give more weight to losses than gains in decision making. This does not necessarily imply irrational behavior.

Thus, humans are extremely risk averse, avoiding losses when gambling with both gains and losses. The value function in Fig. 6.5 illustrates this effect very clearly. The tangent slope of the loss (left side of the graph) is larger than the slope of the gain (right side). This means that the possible loss is felt several times stronger than the gain. On the other hand, in the case of a gamble that could go either way, such as a sure loss or an uncertain but large loss, the slope of the graph in the loss region gradually slows down as risk is pursued. Thus, sensitivity to loss diminishes and becomes risk-seeking.

The above can be summarized as follows:

**Definition 6.3 Human decision making**

- Risk seeking in the area of loss.
- Risk aversion in the area of gain.

We now know that there are two types of utility for humans:

- decision utility: desirability
- experienced utility: a measure of the experience of pleasure or pain

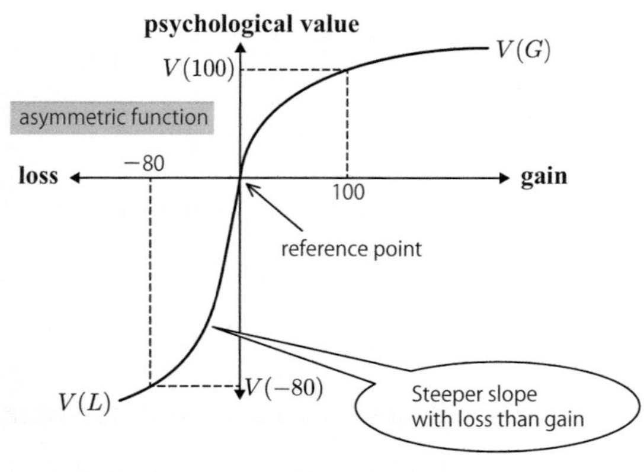


**Figure 6.5:** Kahneman-Tversky value function.

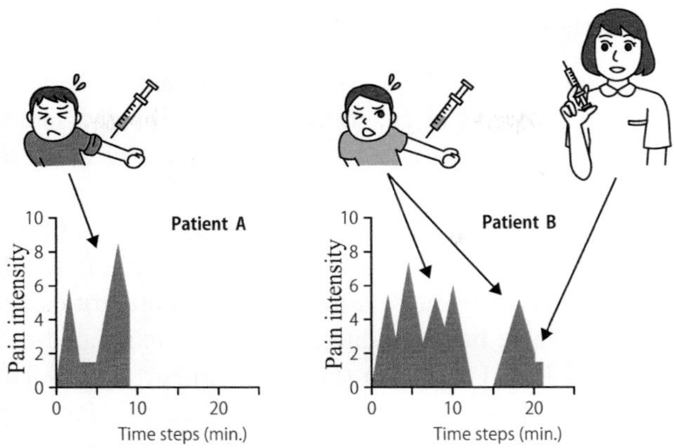


**Figure 6.6:** The peak end rule.

Conventional utility, described in Section 4.1, is decision utility. It explains the reasons for choices. However, as we have seen, decision utility is not always correct in interpreting human behavior. On the other hand, Kahneman focused on experienced utility.

Kahneman describes this utility in terms of extremely painful injections. Suppose that the patient never gets used to this injection and that it hurts the same way every day. Which of the following two is better for the patient?

- Reduce the number of injections from 20 to 18.
- Reduce the number of injections from 6 to 4.

According to classical utility theory, the former reduces the number of injections from 20 to 18, resulting in a 10% reduction. On the other hand, the latter is reduced by 33%. Therefore, the latter is considered to have a greater utility.

Is this true? Something does not seem right. Because it does not take into account the following two points in empirical utility:

- Peak–end law: Evaluation based on memory is determined by the average of pain/pleasure at the peak and at the end.
- Neglect of duration: Duration of pain/pleasure has little influence on the evaluation of total amount.

For this reason, the latter (reducing the number of injections from 6 to 4) is not necessarily better. For example, suppose that two patients A and B feel pain as shown in Fig. 6.6. The duration of pain is shorter for A, but the peak is about the same for both. The pain at the end of the session is significantly less for patient B than for patient A. Therefore, patient B will be less distressed based on memory than patient A.

## 6.6 Let us ask LLMs: How do they respond?

We have discussed prospect theory as a heuristic and bias in human risk perception. Then, do LLMs (and generative AIs) exhibit the same cognitive and perceptual biases? In this section, we will experimentally verify this question.

Specifically, we will ask the same questions to LLMs[5] as we did to humans, using a variety of devices such as prompts, incantations, and CoTs. Then, we observe how LLMs answer the questions and compare the results with human data We will discuss the cognitive differences and similarities between humans and LLMs. Studies have been conducted that have examined the adaptation of prospect theory to LLMs [12]. Such studies are important when applying LLMs to real-world tasks that involve human cognition, e.g., financial engineering and market issues[14].

### 6.6.1 Decoy advertisement

LLMs were asked the same two questions in page 178 times each, and the results are shown in Table 6.3. See Appendix C.1 for examples of typical LLMs' responses. LLMs chose the Web version for Question B because of cost considerations. On the other hand, in Question A, the presence of the decoy advertisement (i.e., Option A-2) made LLMs feel that Option A-3 added value, and they chose that option. The human results are also reproduced in Table 6.3. Comparing the two sets of data, we can see that both humans and LLMs are equally influenced by the decoy advertisement.

**Table 6.3:** Percentage of responses to questions A and B by humans and LLMs.

| | A-1 (%) | A-2 (%) | A-3 (%) | B-1 (%) | B-2 (%) |
|---|---|---|---|---|---|
| Humans | 66.5 | 3.4 | 30.1 | 81.4 | 18.6 |
| ChatGPT | 0.0 | 0.0 | 90.0[6] | 100.0 | 0.0 |

However, LLMs also responded differently from humans. One difference is that LLMs chose only Option A-3 for question A, whereas humans chose Option A-1 most often (66%). This is probably because for LLMs, the emphasis is on a simple superiority-subordination relationship, with Option A-3 being more beneficial than Option A-2. For humans, since the content of the print and Web versions are almost identical, they value the plan with the lower price.

[5]We have used ChatGPT-3.5, ChatGPT-4, and Llama3 as LLMs.

[6]The total does not add up to 100% because 20% of the responses were misguided, such as "it depends on lifestyle."

Another difference is that the values of the alternatives are almost fixed within LLMs. In LLMs, the outcome is almost the same from execution to execution. In contrast, some people even choose Option A-2, i.e., the discarded proposal. This is thought to indicate the diversity of humans, because they do not just perceive superficial merits and demerits, but rather examine the merits of each of their choices.

In addition, we tried the question to the LLMs without the phrase "make the decision as intuitively as possible." In this case, LLMs rarely gave specific alternatives, but only responded by comparing the advantages and disadvantages of each option. In this sense, LLMs are not guiding their answers like human intuition, but rationally understand the correct answer. Conversely, it seems that LLMs dare to include cognitive bias when they answer as humans do.

### *6.6.2 The paradox of Allais*

We questioned the choice of jars shown in Fig. 6.1. As noted in 183 page, many humans choose jars A and D, contrary to the expected utility theory.

Simply asking LLMs, "Choose the more economically rational jar," they chose jars B and D. This is a choice that does not violate the expected utility theory, and LLMs responded, "Calculate the expected value." Furthermore, the same answer was given when LLMs were asked to think in a step-by-step manner without using CoT. In other words, LLMs did not think intuitively like humans, but always answered rationally using mathematical formulas. In particular, since LLMs do not have human-like value functions (i.e., Kahneman-Tversky function) in their thinking process, the phenomenon of loss aversion does not occur, and they always adopt the one with the larger expected value.

Next, we asked LLMs, "Which would you expect the average person to choose for each?" LLMs answered as follows:

> We would expect the average person to choose jar A (a sure low amount) and jar C (a relatively high probability of a low amount). This is based on the psychological tendency to avoid the risk that jars B and D, with their potential for higher payouts, have.

The above results confirm that LLMs know and can consider the principle of loss aversion. When asked the question with the prompt such as "Do not think about the expected value," they chose jar A because of the certainty of a profit and jar C because of the probability of a small profit.

### 6.6.3 How much bet would LLMs turn down?

Consider the amount of money you would decline to bet shown in Table 6.2. We have asked LLMs the following question:

> Consider the following bet: You have a 50% chance of losing $100 but a 50% chance of getting ( A ) dollars. What is the minimum value of A when you decline this bet? Please never give a vague answer, and answer with a number, such as $xxx.

LLMs' responses followed the classical utility theory. The following is an example of their responses.

> The minimum value of A is $100. If there is a 50% probability that the amount you lose exceeds the amount you gain, it is rational to take the bet.

In addition, when asked questions using prompts that implied that humans and LLMs have different cognitive characteristics (e.g., "Please answer the question taking into account human risk perception characteristics."), LLMs gave the same responses.

> It is known that when one considers the Kahneman-Tversky value function, one finds a hypersensitivity to loss. People tend to avoid losses. In the case of this bet, the choice depends on how much can be gained for the $100 that can be lost. Since most people are loss averse, they will decline the bet unless the gain is more than $100 they could lose. In other words, the minimum value of A is considered to be greater than $100. A numerical answer would be in the form of a minimum value for A greater than $100, e.g., "$150."

These results indicate that LLMs follow the classical utility theory by calculating expectations. On the other hand, LLMs understand the Kahneman-Tversky value function and the concept of loss aversion. Thus, it may be possible to let them reason on that basis.

### 6.6.4 Endowment effect

If examples such as shown in Figs. 6.2 and 6.3 were entered directly into LLMs, they would judge that it was a thought experiment about the prospect theory and begin explaining the theory. Therefore, in order to examine the opinions of the LLMs themselves, we used a prompt that created another setting of college students and asked LLMs to answer as a line. An example of that question-and-answer session is shown in Appendix C.2. The results show that LLMs fall into the endowment effect and estimate the value of their possessions (i.e., pens) at a higher level. However, in this situation, the seller knows that there is demand for the item, so he can set the price from a higher position. In this sense, the psychological perspective may not be the only determinant of the price difference. When we actually asked about the price difference, LLMs did not mention the content of the endowment effect. Therefore, it is conceivable that LLMs behaved in a human-like manner in this experiment.

Next, to eliminate the supply and demand factor, we asked the same questions as in Figs. 6.2 and 6.3 about a backpack bought 5 years ago for 10,000JPY (**Prompt1**).[7] The results of the five buying and selling questions while initializing the chat are shown in Table 6.4. The average purchase and sale prices matched at 3,800JPY for both. Furthermore, a reversal between the selling price and the purchase price also occurred. Therefore, no endowment effect occurred in this case, and the price was objectively determined from the product. Another question was asked with the time of purchase changed to a more recent date (**Prompt2**). The averages of the sale and purchase prices were 7,900JPY and 6,500JPY, respectively, confirming that the sale price was predominantly higher and that the endowment effect was occurring.

**Table 6.4:** LLMs response results.

| | **Prompt 1** | | **Prompt 2** | |
|---|---|---|---|---|
| | **Selling price** | **Purchase price** | **Selling price** | **Purchase price** |
| response1 | 4,500 | 3,000 | 8,000 | 6,000 |
| response2 | 3,000 | 5,000 | 8,000 | 7,000 |
| response3 | 4,500 | 3,500 | 7,500 | 6,500 |
| response4 | 4,000 | 3,000 | 8,000 | 6,000 |
| response5 | 3,000 | 4,500 | 8,000 | 7,000 |
| **average** | 3,800 | 3,800 | 7,900 | 6,500 |

[7]The exchange rate at the time of the question to LLMs was approximately 140 Japanese yen (JPY) to the dollar.

When we asked LLMs about the price difference in **Prompt2**, they responded with the following four effects:

1. Opportunity cost and psychological loss
   The seller has already paid 10,000JPY for the backpack and is sensitive to the amount he can recover by selling it. Therefore, the seller has a psychological urge to sell it at a price close to the purchase price. On the other hand, the buyer has not yet paid for the backpack and wants to purchase it at the lowest possible price.
2. Asymmetric information
   Buyers cannot fully confirm the condition of an item. Therefore, even if the item is like new, the risk that it may have hidden defects is taken into account when determining the price. On the other hand, sellers often set a higher price because they know.
3. Room for negotiation
   Sellers typically set a slightly higher price to leave room for negotiation in the price. Buyers may offer lower than desired price in anticipation of price negotiation.
4. Psychological ownership effect (endowment effect)
   People tend to perceive the value of things they own to be higher than their actual market value. Therefore, sellers may perceive a higher value for their backpacks than buyers do for their own backpacks.

Since the first and forth points are the content of prospect theory, it is clear that LLMs understand the theory and determine their behavior accordingly. The second and third points are those that come with human sales and will have a non-eliminable influence on LLMs to behave in a human-like manner.

By specifying the customer's situation, LLMs can reflect some of the psychological conditions specific to that situation. This indicates that LLMs may be able to imagine the demand by defining the target. For example, LLMs could be used for marketing and other applications, such as for planning measures to reach potential customers.

### 6.6.5 Framing effects

Framing effect is a phenomenon in which humans' impression of the same meaning can be greatly changed by simply changing the way it is expressed. Then, does the framing effect occur in LLMs? To find out, let us ask LLMs the same two questions in 188 page. Recall that in the above two questions the first and next options are equivalent, just with different expressions (i.e., “save” v.s. “die”). In other words, if one chooses Option1 in the first question, it seems reasonable to choose Optoin3 in the next question. However, in humans, choosing Option1 and Option4 was predominantly more common. This is because, due to cognitive bias, an option with the expression “die” representing a loss is more likely to be chosen than with the expression “save” representing a gain.

Now let us ask LLMs these questions. A specific prompt[8] is given in the input. Example prompts are shown in Appendix C.3

When appropriate responses were received, option 2 was often chosen. A sample response is shown in **Output1** in Appendix C.3. LLMs seem to be hopeful about the possibility of remaining alive. For the second question, they chose option 3. An example response is shown in **Output2** in Appendix C.3. In this case, the emphasis is on avoiding the risk of dying for all. From this result, we can see that humans choose the option that contains the word “death” as a loss. On the other hand, LLMs are more likely to choose the option that represents a gain, i.e., “to be saved.” Therefore, LLMs do not necessarily have the loss aversion tendency that humans have.

As a next experiment, we ask LLMs the following question:

**Question**
If you went to an amusement park and were offered the following two attractions, which one would you prefer to ride?

**Option A** The attraction with a 95% satisfaction rate.

**Option B** The attraction with which 1 in 20 people would not be satisfied.

Both options A and B are equally satisfied. The only difference is that A is positive and B is negatively written. When LLMs were asked this question, they chose option A as shown in **Output3** in Appendix C.3. After repeating the above question 10 times, LLMs continued to choose option A. This result shows that LLMs will choose options with positive content if they have the same level of satisfaction.

[8]LLMs often mentioned ethical issues or replied that the decision would be subjective. Thus, we have entered more conditions beforehand, e.g., “the decision should be made by an individual.”

Now let us take this to an even more extreme level. In the next question, we have taken the satisfaction level of Option B to the extreme, so that a person would normally consider Option B to be the attraction of choice.

**Question**
If you went to an amusement park and were offered the following two attractions, which one would you prefer to ride?

**Option A** The attraction with a 95% satisfaction rate.

**Option B** The attraction with which 1 in 100 million people would not be satisfied.

The results of asking LLMs this question are shown in **Output4** in Appendix C.3. As can be seen, LLMs do not make their choices based on the level of satisfaction with an option, but rather on the degree of positivity of the option (in this case, the level of satisfaction).

We further tested how much lower the degree of satisfaction would lead LLMs to choose Option B (i.e., 1 out of 20 unsatisfactory attractions). When LLMs were asked 10 times with different satisfaction levels for Option A, the number of times they chose A and B respectively is shown in Table 6.5.

**Table 6.5:** Changes in LLMs' response results for different levels of satisfaction with option A.

| **Satisfaction with option A (%)** | **Number of times** | |
|---|---|---|
| | **A was chosen** | **B was chosen** |
| 95 | 10 | 0 |
| 75 | 10 | 0 |
| 65 | 9 | 1 |
| 60 | 1 | 9 |
| 50 | 1 | 9 |

The results show that the number of times A and B are chosen is reversed between 65% and 60% satisfaction. The LLMs' responses to the 65% and 60% satisfaction questions are shown in **Output5** and **Output6** in Appendix C.3, respectively. Let us note the following characteristic statements in each of the responses:

**65%** Option A is generally likely to be satisfied by more than half.

**60%** Since the satisfaction rate for Option A is low at 60%, $\cdots$

In other words, the criterion for choosing between options A and B depends on whether A has a high or low satisfaction rating. As a result, it was confirmed that LLMs learn to evaluate low satisfaction levels of 60% or lower and high satisfaction levels of 65% or higher.

These results indicate that LLMs have greater cognitive and perceptual biases than humans. When LLMs are applied to the real world, there is a risk that they will not make suggestions that are originally optimal because of their cognitive bias. When cognitive biases such as the framing effect are involved, it is difficult for LLMs to make decisions similar to those made by humans because it is difficult to recognize the correlation between the biases.

### *6.6.6 Reference points*

Consider **Case 1** and **Case 2** of page 180. Remember that in **Case 1**, the reference point is $1,000 and there is a possibility of gaining new money from it. In **Case 2**, the reference point is set at $2,000 and there is a possibility of losing money.

The results of asking LLMs 10 times each case are shown in Table 6.6. Options A and B mean the same thing in both cases, just with different reference points. In other words, Option A has the same probability of receiving $1,000 or $2,000, while Option B will certainly get $1,500. Therefore, for a rational agent, the answers in the two cases should be the same. However, as can be seen from the table, in **Case 1**, where the reference point is $1,000, option B is chosen more often, and in **Case 2**, where the reference point is $2,000, option A is chosen more often. This means that when the reference point is low and money increases, certainty is prioritized. On the other hand, when the reference point is high and money decreases, the certainty of money decreasing is disliked, and gambling is tolerated. In other words, as in humans, a cognitive error caused by the reference point occurs in LLMs.

**Table 6.6:** LLMs' response results for reference points.

| Option | Case 1 | Case 2 |
|---|---|---|
| A | 2 | 8 |
| B | 7 | 1 |
| Otherwise[9] | 1 | 1 |

Note that the question to LLMs on this question included the phrase "answer intuitively." As a result, LLMs were set to answer as if they themselves were human, which may reflect the irrational cognitive bias assumed in the prospect theory. When the question was asked without the phrase "answer intuitively,"

[9]Other cases included responses that vary by situation and values.

LLMs made decisions based on their expectations and gave answers that did not explicitly state which they would choose.

### 6.6.7 Summary of psychologically biased results

Previous experimental results show similarities and differences between humans and LLMs in cognitive biases. It should be noted, however, that these results are not always consistent.

For instance, when we included the phrase "answer intuitively" in the questions to the LLMs on the reference point question, LLMs responded as humans, reflecting irrational cognitive biases assumed in prospect theory. When the question was asked without the phrase "answer intuitively," LLMs made rational judgments based on their expectations. On the other hand, responses to the question about decoy ads did not change regardless of whether or not the phrase "answer intuitively" was used. This indicates that LLMs may or may not incorporate cognitive biases in the learning process. Therefore, when applying LLMs to real-world situations, it would be important to specify more precisely the position from which LLMs respond and to clarify the extent to which LLMs understand the theory behind the question.

Asking LLMs about trends in human judgment does not mean that LLMs can make judgments with the same biases and rationales as humans. Therefore, it is necessary to be fully aware of the possibility of inaccurate results. Furthermore, it may be difficult to use LLMs results for questions involving irrational psychological effects in humans.

In the case of questions where a number is given in the prompt, LLMs do not think intuitively as humans do, but tend to answer rationally, using mathematical formulas and expected values. As a result, while human-like cognitive errors did not occur, LLMs did not decide which is better when the expected values are equal, and sometimes responded that it depends on the individual utility function.

# Chapter 7

# Conclusion

> All life is problem solving. From the amoeba to Einstein, the growth of knowledge is always the same. (Karl Popper)

## 7.1 Summary and concluding remarks

We have described recent developments and potential applications for generative AI, focusing on emergence and evolutionary methods. We provided the recent progress of researches on LLMs by mean of swarm and evolutionary intelligence. We also explained the differences in thinking between LLMs and humans and gave empirical evidences.

Based on the previous experimental results, appropriate judgments at human users' places are required for LLMs to be properly utilized in the real world. LLMs could be a useful tool that allows us to think in a reasonable manner. However, we have also found that LLMs and humans have different cognitive biases. For instance, in consideration of the framing and endowment effects, the use of LLMs would be quite different when applying to the following real-world tasks:

- Financial engineering field, which is more averse to losses than to taking risks for the sake of profits.
- Medical field where the main purpose is to save people.

In such cases, if LLMs are given prompts to think only in terms of expected value, they will not necessarily be effective because of the divergence from human thinking.

As mentioned in Section 1.2, one of the blind spots of LLMs is that they are not good at arithmetics or mathematics. LLMs are "probabilistic parrots" so that they can gets simple math problems wrong.

On the other hand, humans are good at mathematics and abstract thinking. Studies have shown that some degree of mathematical and causal reasoning is innate in humans and animals. For example, very simple addition and subtraction can be done by infants [167].

In this experiment, the baby watches the following play:

1. Place a stuffed animal on the stage (Fig. 7.1(a)).

2. Hide the stuffed animal with a screen (Fig. 7.1(b)).

3. Show another stuffed animal while keeping the first stuffed animal hidden by the screen.

4. Put this stuffed animal behind the screen so that only the screen is visible on the stage (Fig. 7.1(c)).

5. Remove the screen.

At this point, one would naturally expect to see two stuffed animals. However, after the screen was removed, the baby's reaction was different between when there were two stuffed animals and when there was one. When the baby saw two (Fig. 7.1(d)), it did not pay much attention to the screen and looked elsewhere (Fig. 7.1(e)). When the baby unexpectedly saw one stuffed animal, the baby was surprised and looked at that for a longer period of time (Fig. 7.1(f)). Similarly, when one was taken from the two stuffed animals, the baby also looked longer when there were two. Even Asian elephants have been reported to be able to perform such simple addition. On the other hand, birds are known to lack this ability [28]. Suppose, for example, that two people enter a cave and then one person leaves. At this time, the bird approaches, thinking that it is now safe. This is a technique known to hunters. Why do human babies and elephants instinctively possess certain numerical cognitive abilities? Perhaps the acquisition of such abilities contributed to their survival during the evolutionary process. In the previous cave, we saw a large number of these animals. Imagine if a human were in a cave watching lions come and go. Similarly, there is evidence that babies understand the basics of physics (e.g., a cup does not suddenly float up). This is called "genetic memory,'' which is the genetic transmission of knowledge and skills along with instincts and behaviors.

In this sense, some learning is difficult to explain with neural networks alone [107]. Rats learn to avoid nauseating foods after many hours. Such learning is acquired in a single opportunity, and once it is acquired, it does not disappear. If this learning had occurred in a neural network (i.e., error back-propagation), the rat would have died and become extinct. Learning avoidance from a one-time

**Figure 7.1:** 1+1=2?

experience in rats is called "prepared learning." This is in contrast to the characteristics of deep learning. Nausea conditioning is also influenced by the unique evolutionary history of the animal [64]. Rats, for example, have poor eyesight and rely on their sense of taste so as to smell to find food at night. Thus, they quickly develop an aversion to strange tasting foods, but they do not learn to dislike foods that look strange. In quail, on the other hand, the learning pattern is quite different, and aversion is more easily associated with color. One explanation is that intrinsic stimuli may be responsible for motivating learning.

We can recognize the presence of a snake before we are even aware that we are seeing it [125]. This phenomenon is called "preconscious detection." When an image of a snake appears on a computer screen so fast that we do not actually notice it, we feel physiologically disturbed. But such reactions do not occur for non-threatening objects such as mushrooms or flowers.

Why do mothers talk to their newborn babies? It is not just for learning, but to promote the morphological development of the baby's brain. It is thought that we have acquired this clever way of building up the brain during our evolution. Many neurons in the brain combine randomly, eliminating the need for elaborate blueprints. Randomly connected regions are organized as the baby interacts with its environment. Moreover, nature takes advantage of modularity to reuse useful parts again as they are found. This can lead to rapid growth in intelligence with only a few genetic variations.

For these reasons, it is plausible to think that mathematical thinking and logical reasoning are products of evolution. How generative AI and LLMs will change this idea is an interesting topic for future research.

As an extension of this book, there has been a recent flurry of researches on "Large Language Models as evolutionary computation [48] (see Section 5.4.9 for instance)."

These topics were not discussed in this book because they are in the process of development. We hope for further research progress on generative AI with the integration of swam and evolutionary intelligence.

# Appendix A

# Software Packages

We have made available software implementations of many of the evolutionary or swarm systems described in this book. We hope that this software will give the readers some hands-on experience with evolutionary systems, in addition to the theoretical knowledge provided by the previous chapters. We suggest that the readers explore the software while reading each chapter, trying new parameters and data sets. Doing this you will get a clearer understanding of the possibilities and limits of the presented systems.

In this appendix we give general instructions for the use of each available software package. The packages themselves can be downloaded from the Iba Laboratory's web page: http://www.iba.t.u-tokyo.ac.jp/english/.

Please follow the download instructions available in that webpage. Besides the software packages, the webpage also contains user manuals and bulletin boards. Please feel free to use those in case you need assistance using the software.

Please be aware that the software packages available in the above webpage are copyrighted to the authors. The programs may be altered, added to, or removed without notice. We accept no liability for any damages resulting from the use of the programs, or of the output generated by executing them. A more detailed license accompanies each software package.

## A.1 TSP solver by ACO

This software visualizes the process of solving the Traveling Salesman Problem (TSP) using Ant Colony Optimization (ACO). The main feature of this software is that you can observe how the concentration of pheromone changes as you

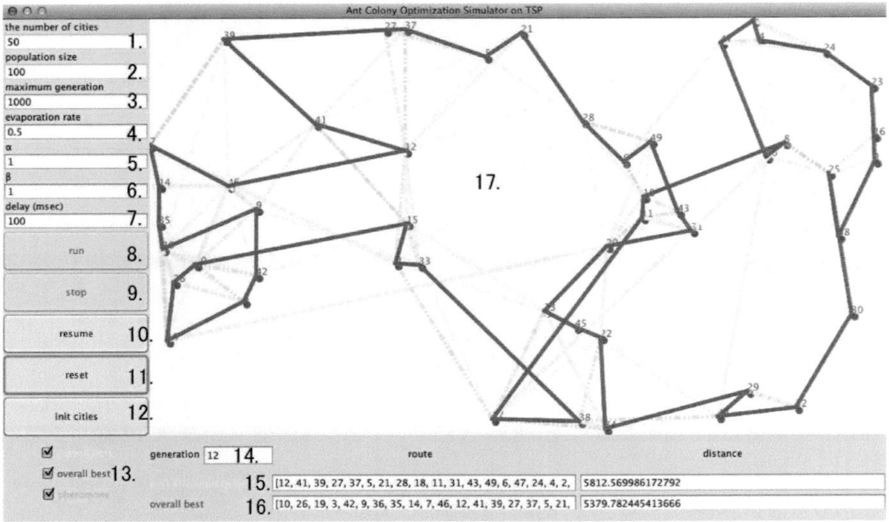


**Figure A.1:** TSP Simulator by ACO.

move the city during the execution. To run this software, JRE must be installed on your computer.

The usage is as follows. Figure A.1 is a screenshot of the simulator when it is run. In this simulator, pheromone trails are indicated by yellow-green lines. The darker the line, the higher the pheromone concentration. The shortest paths found at the moment are shown in red.

The command inputs and outputs displayed on the screen are described below. The numbers match the markings in the figure.

1. the number of cities: Enter the number of cities.

2. population size: Enter the number of individuals, i.e., the number of ants per generation.

3. maximum generation: Enter the maximum number of generations to be trained.

4. evaporation rate :Enter the pheromone evaporation rate in the range of 0.0 to 1.0. If the value is less than or equal to 0.0, it is set to be 0.0, and if the value is greater than 1.0, it is set to be 1.0. The larger the value, the more pheromone information is carried over to the next generation.

5. $\alpha$: Priority of the pheromone information. The selection probability of a city is proportional to the $\alpha$ power of the pheromone value.

6. $\beta$: Priority of the heuristic information. The probability of selecting a city is proportional to $\beta$ power of the heuristic (inverse of distance).

7. *DELAY*: Delay per generation. Lowering this value speeds up the drawing, but increases computational costs.
8. run button: Clicking this button starts the ACO.
9. stop button: Click this button to stop ACO temporarily. Some parameters can be changed while the ACO is stopped.
10. resume button: Click this button to resume the ACO while it is paused.
11. reset button: Clicking on this button resets the ACO while leaving the city layout as it is.
12. init. cities button: Resets all information and rearranges the cities randomly.
13. checkbox: Only the checkboxes checked here among current best, overall best, and pheromone will be drawn in the drawing field.
14. generation: Displays the number of current generations.
15. best at current generation: Displays the route and distance of the best individual in the current generation.
16. overall best: Displays the route and distance of the best individual.
17. Drawing field: The numbered cities, current best (best individuals of the current generation, yellow), overall best (best individuals, red) and pheromone (pheromone, green) are displayed here. The pheromone is displayed in particular as a darker color when the pheromone density is high and a lighter color when the density is low. Cities can be dragged and moved at any time.

## A.2 Multi-objective optimization by GA

To test and experiment with evolutionary multi-objective optimization (EMO; see Section 4.3), a Pareto GA simulator is available. This is based on VEGA (see 126 page).

The program execution screen of the simulator is shown in Fig. A.2. Each red point is an individual, and the number of dominant individuals relative to that point (including itself) is shown nearby. For an individual on the Pareto front, the number is 1.

Let us use this simulator to minimize the value of two objectives (fitness functions) $f_1(x)$ and $f_2(x)$. The fitness functions can be entered in the "Function 1" and "Function 2" boxes. After setting the GA parameters (for example, population size and generation number), click the "Start" button to begin the search.

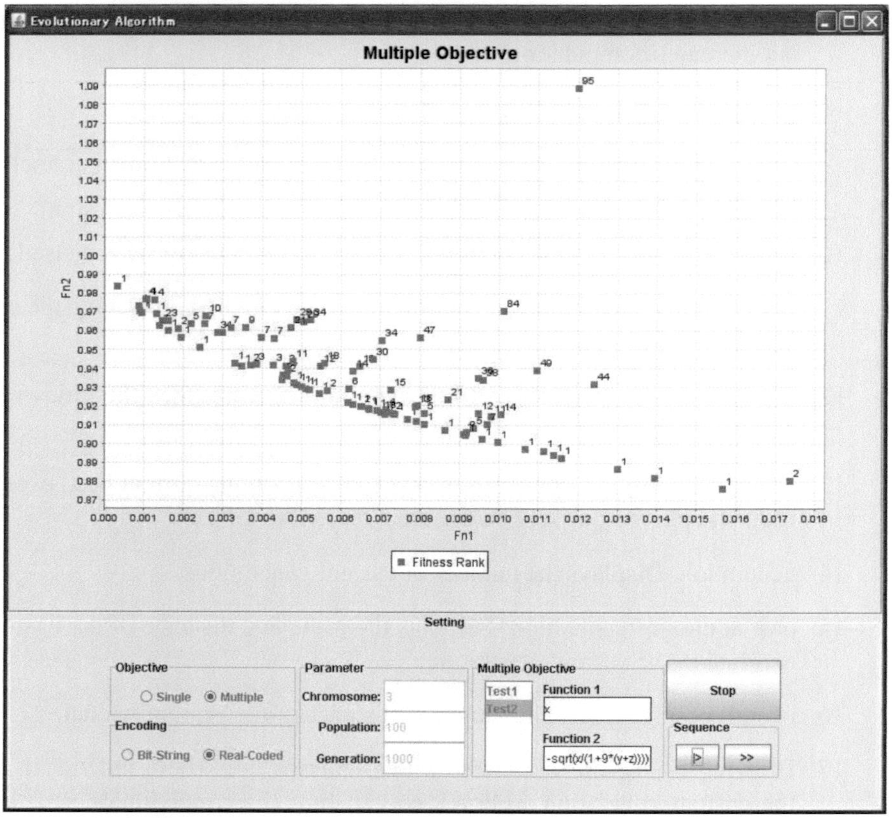


**Figure A.2:** Pareto optimization simulator.

During execution, autozoom can be disengaged by clicking the graph and restored by clicking it again.

For the multi-objective optimization shown in Figs. 4.19 and 4.20, let us enter the corresponding functions directly into the function boxes. For instance, the fitness function for $\alpha = 0.5$ is written as follows:

```
Function 1:  x
Function 2:  1-sqrt(x)
```

And the fitness function for $\alpha = 2.0$ is written as

```
Function 1:  x
Function 2:  1-x*x
```

To optimize the functions shown in Fig. **??**, enter the following in their respective boxes:

```
Function 1:  x
Function 2:  1-pow(x,0.25)-x*sin(10*3.1415*x)
```

**Table A.1:** List of available functions.

| | Symbol | Description |
|---|---|---|
| **Operators** | $+ - * /$ | Four arithmetic operations |
| | $\wedge$ | Power function |
| | $-$ | Negation |
| | ! | Factorial |
| **Functions** | abs(x) | Absolute function |
| | pi() | $\pi$(3.14159...) |
| | degrees(x) | Convert to angle |
| | round(x,0) | Rounding function |
| | fact(x) | Factorial function |
| | sqrt(x) | Square root function |
| | exp(x) | Exponential function |
| | log(x,y) | Logarithm function (base is y) |
| | sin(x) | Sine function |
| | cos(x) | Cosine Function |
| | tan(x) | Tangent function |
| | asin(x) | Arcsine function |
| | acos(x) | Arccosine function |
| | atan(x) | Arctangent function |
| | sinh(x) | Hyperbolic sine funcion |
| | cosh(x) | Hyperbolic cosine function |
| | tanh(x) | Hyperbolic tangent function |
| | rand() | Random number from 0 to 1 |
| | Gauss(m,s) | Gaussian random number (mean=m, variance=s) |
| | Max(x,y) | A maximum value of x and y |
| | Min(x,y) | A minimum value of x and y |
| | if(bool,x,y) | Return $x$ if bool is **true**, otherwise **false** |

After entering the functions, execute the simulation, and observe the complex Pareto front conformation.

You can use transcendental as well as arithmetic functions to define the fitness functions on the simulator. Table A.1 list the functions that can be used. The input expression is a three-variable function, in $x, y$, and $z$. For each variable a range of possible values can be established as any two positive numbers. On the simulator, the Pareto curve is displayed on a graph with $f_1(x)$ and $f_2(x)$ being shown on the horizontal and vertical axes, respectively.

# *Appendix B*

# Source Codes Generated by LLMs

This appendix contains the raw python codes generated by LLMs. These can be executed with very little human intervention. Of course, the code generated by LLMs will be different each time, even if the prompt is the same. The codes shown here are a typical output example.

## B.1 Limited evaluation swarm optimizer

```
import numpy as np
import matplotlib.pyplot as plt

class LESO:
    def __init__(self, objective_function, initial_population,
         max_iterations, max_eval_count):
        self.objective_function = objective_function
        self.population = initial_population
        self.max_iterations = max_iterations
        self.max_eval_count = max_eval_count
        self.eval_count = 0

    def evaluate_fitness(self, x):
        return self.objective_function(x)

    def optimize(self):
        for iteration in range(self.max_iterations):
            selected_solutions = self.select_solutions()
            for solution in selected_solutions:
```

```
                fitness = self.evaluate_fitness(solution)
                self.eval_count += 1
                if self.eval_count >= self.max_eval_count:
                    return self.population[np.argmin([self.evaluate_fitness(sol) for sol in self.population])]
            self.update_population(selected_solutions)
        return self.population[np.argmin([self.evaluate_fitness(sol) for sol in self.population])]

    def select_solutions(self):
        # Placeholder for adaptive selection strategy
        return self.population[:len(self.population) // 2] # Select half of the population for evaluation

    def update_population(self, selected_solutions):
        # Placeholder for adaptive swarm dynamics update
        for i in range(len(selected_solutions)):
            self.population[i] = self.explore(selected_solutions[i])

    def explore(self, solution):
        # Placeholder for exploration strategy
        return solution + np.random.normal(0, 0.1, size=solution.shape) # Add random noise to the solution
```

## B.2 Enhanced swarm exploration and exploitation optimizer

```
import numpy as np
from sklearn.gaussian_process import GaussianProcessRegressor
from sklearn.gaussian_process.kernels import RBF
from scipy.optimize import minimize
import matplotlib.pyplot as plt

class ESEEO:
    def __init__(self, objective_function, initial_population, max_iterations, max_eval_count):
        self.objective_function = objective_function
        self.population = initial_population
        self.max_iterations = max_iterations
        self.max_eval_count = max_eval_count
        self.eval_count = 0
        self.surrogate_model = GaussianProcessRegressor(kernel=RBF(length_scale=1.0), alpha=0.1)

    def evaluate_fitness(self, x):
```

```
        return self.objective_function(x)

    def surrogate_evaluate_fitness(self, x):
        return self.surrogate_model.predict(np.atleast_2d(x))[0]

    def optimize(self):
        for iteration in range(self.max_iterations):
            selected_solutions = self.select_solutions()
            for solution in selected_solutions:
                fitness = self.evaluate_fitness(solution)
                self.eval_count += 1
                self.update_surrogate_model(np.atleast_2d(solution), fitness)
                if self.eval_count >= self.max_eval_count:
                    return self.population[np.argmin([self.surrogate_evaluate_fitness(sol) for sol in self.population])]
            self.update_population(selected_solutions)
        return self.population[np.argmin([self.surrogate_evaluate_fitness(sol) for sol in self.population])]

    def select_solutions(self):
        # Placeholder for adaptive selection strategy
        return self.population[:len(self.population) // 2] # Select half of the population for evaluation

    def update_surrogate_model(self, x, y):
        self.surrogate_model.fit(x, [y])

    def update_population(self, selected_solutions):
        # Placeholder for adaptive exploration and exploitation strategies
        for i in range(len(selected_solutions)):
            self.population[i] = self.explore(selected_solutions[i])

    def explore(self, solution):
        # Placeholder for exploration strategy
        return solution + np.random.normal(0, 0.1, size=solution.shape) # Add random noise to the solution
```

# B.3 Diversity-enhanced hybrid optimization

```
import numpy as np
import matplotlib.pyplot as plt
class DEHO:
```

```
    def __init__(self, objective_function, max_iterations, population_size, search_space_dimension, search_space_boundaries):
        self.objective_function = objective_function
        self.max_iterations = max_iterations
        self.population_size = population_size
        self.search_space_dimension = search_space_dimension
        self.search_space_boundaries = search_space_boundaries
        self.population = None
        self.fitness_values = None

    def initialize_population(self):
        self.population=np.random.uniform(low=self.search_space_boundaries[0],high=self.search_space_boundaries[1],size=(self.population_size,self.search_space_dimension))

    def evaluate_fitness(self):
        self.fitness_values = np.apply_along_axis(self.objective_function, 1, self.population)

    def waggle_dance_communication(self):
        # Calculate fitness rank of each solution
        fitness_rank = np.argsort(self.fitness_values)

        # Calculate waggle dance information based on fitness rank
        waggle_dance_info = 1 / (fitness_rank + 1) # Inverse rank (higher fitness => lower rank => more information)

        # Share waggle dance information among solutions
        for i, solution in enumerate(self.population):
            self.population[i] += np.random.uniform(-0.1, 0.1, size=solution.shape) * waggle_dance_info[i]

    def update_velocities(self, inertia_weight=0.5, cognitive_weight=0.5, social_weight=0.5):
        personal_best = self.population[np.argmin(self.fitness_values)]
        global_best = personal_best

        for i, solution in enumerate(self.population):
            velocity = np.random.uniform(-0.1, 0.1, size=solution.shape) # Initialize velocity randomly
            velocity += inertia_weight * velocity # Inertia component
            velocity += cognitive_weight * np.random.uniform() * (personal_best - solution) # Cognitive component
```

```
            velocity += social_weight * np.random.uniform() * (global_best - solution) # Social component
            self.population[i] += velocity

    def perform_levy_flights(self, levy_flight_probability=0.1, levy_flight_scale=0.1):
        for i, solution in enumerate(self.population):
            if np.random.uniform() < levy_flight_probability:
                levy_flight_step = levy_flight_scale * np.random.standard_cauchy(size=solution.shape)
                self.population[i] += levy_flight_step

    def optional_local_search(self):
        # Optionally apply local search to improve solutions
        for i, solution in enumerate(self.population):
            improved_solution = self.hill_climbing(solution)
            if self.objective_function(improved_solution) < self.objective_function(solution):
                self.population[i] = improved_solution

    def hill_climbing(self, solution, step_size=0.1, max_iterations=10):
        current_solution = solution.copy()
        for _ in range(max_iterations):
            # Generate a random neighboring solution
            neighbor_solution = current_solution + np.random.uniform(-step_size, step_size, size=current_solution.shape)
            # If the neighbor has better fitness, move to it
            if self.objective_function(neighbor_solution) < self.objective_function(current_solution):
                current_solution = neighbor_solution
        return current_solution

    def run(self):
        self.initialize_population()
        self.evaluate_fitness()

        for iteration in range(self.max_iterations):
            self.waggle_dance_communication()
            self.update_velocities()
            self.perform_levy_flights()
            self.optional_local_search()
            self.evaluate_fitness()

        best_solution_index = np.argmin(self.fitness_values)
        best_solution = self.population[best_solution_index]
        best_fitness = self.fitness_values[best_solution_index]

        return best_solution, best_fitness
```

## B.4 Merge heuristics

```
import numpy as np
import matplotlib.pyplot as plt

class Particle:
    def __init__(self, dimension, lower_bound, upper_bound):
        self.position = np.random.uniform(lower_bound, upper_bound, size=dimension)
        self.velocity = np.random.uniform(-1, 1, size=dimension)
        self.best_position = self.position
        self.best_fitness = float('inf')

class PSO:
    def __init__(self, objective_function, dimension, num_particles, lower_bound, upper_bound, max_iterations, inertia_weight, cognitive_weight, social_weight):
        self.objective_function = objective_function
        self.dimension = dimension
        self.num_particles = num_particles
        self.lower_bound = lower_bound
        self.upper_bound = upper_bound
        self.max_iterations = max_iterations
        self.inertia_weight = inertia_weight
        self.cognitive_weight = cognitive_weight
        self.social_weight = social_weight
        self.global_best_position = None
        self.global_best_fitness = float('inf')
        self.particles = [Particle(dimension, lower_bound, upper_bound) for _ in range(num_particles)]

    def optimize(self):
        for iteration in range(self.max_iterations):
            for particle in self.particles:
                fitness = self.objective_function(particle.position)
                if fitness < particle.best_fitness:
                    particle.best_fitness = fitness
                    particle.best_position = particle.position.copy()
                if fitness < self.global_best_fitness:
                    self.global_best_fitness = fitness
                    self.global_best_position = particle.position.copy()
            for particle in self.particles:
                r1 = np.random.random(self.dimension)
                r2 = np.random.random(self.dimension)
                cognitive_component = self.cognitive_weight * r1 * (particle.best_position - particle.position)
```

```
                social_component = self.social_weight * r2 * (self.global_best_position - particle.position)
                particle.velocity = self.inertia_weight * particle.velocity + cognitive_component + social_component
                particle.position += particle.velocity
                particle.position = np.clip(particle.position, self.lower_bound, self.upper_bound)
        return self.global_best_position, self.global_best_fitness

class DEHO:
    def __init__(self, objective_function, max_iterations, population_size, search_space_dimension, search_space_boundaries):
        self.objective_function = objective_function
        self.max_iterations = max_iterations
        self.population_size = population_size
        self.search_space_dimension = search_space_dimension
        self.search_space_boundaries = search_space_boundaries
        self.population = None
        self.fitness_values = None

    def initialize_population(self):
        self.population=np.random.uniform(low=self.search_space_boundaries[0],high=self.search_space_boundaries[1],size=(self.population_size,self.search_space_dimension))

    def evaluate_fitness(self):
        self.fitness_values = np.apply_along_axis(self.objective_function, 1, self.population)

    def waggle_dance_communication(self):
        # Calculate fitness rank of each solution
        fitness_rank = np.argsort(self.fitness_values)

        # Calculate waggle dance information based on fitness rank
        waggle_dance_info = 1 / (fitness_rank + 1) # Inverse rank (higher fitness => lower rank => more information)

        # Share waggle dance information among solutions
        for i, solution in enumerate(self.population):
            self.population[i] += np.random.uniform(-0.1, 0.1, size=solution.shape) * waggle_dance_info[i]

    def update_velocities(self, inertia_weight=0.5, cognitive_weight=0.5, social_weight=0.5):
```

```
        personal_best = self.population[np.argmin(self.fitness_values)]
        global_best = personal_best

        for i, solution in enumerate(self.population):
            velocity = np.random.uniform(-0.1, 0.1, size=solution.shape) # Initialize velocity randomly
            velocity += inertia_weight * velocity # Inertia component
            velocity += cognitive_weight * np.random.uniform() * (personal_best - solution) # Cognitive component
            velocity += social_weight * np.random.uniform() * (global_best - solution) # Social component
            self.population[i] += velocity

    def perform_levy_flights(self, levy_flight_probability=0.1, levy_flight_scale=0.1):
        for i, solution in enumerate(self.population):
            if np.random.uniform() < levy_flight_probability:
                levy_flight_step = levy_flight_scale * np.random.standard_cauchy(size=solution.shape)
                self.population[i] += levy_flight_step

    def optional_local_search(self):
        # Optionally apply local search to improve solutions
        for i, solution in enumerate(self.population):
            improved_solution = self.hill_climbing(solution)
            if self.objective_function(improved_solution) < self.objective_function(solution):
                self.population[i] = improved_solution

    def hill_climbing(self, solution, step_size=0.1, max_iterations=10):
        current_solution = solution.copy()
        for _ in range(max_iterations):
            # Generate a random neighboring solution
            neighbor_solution = current_solution + np.random.uniform(-step_size, step_size, size=current_solution.shape)
            # If the neighbor has better fitness, move to it
            if self.objective_function(neighbor_solution) < self.objective_function(current_solution):
                current_solution = neighbor_solution
        return current_solution

    def run(self):
        self.initialize_population()
        self.evaluate_fitness()

        for iteration in range(self.max_iterations):
```

```
            self.waggle_dance_communication()
            self.update_velocities()
            self.perform_levy_flights()
            self.optional_local_search()
            self.evaluate_fitness()

        best_solution_index = np.argmin(self.fitness_values)
        best_solution = self.population[best_solution_index]
        best_fitness = self.fitness_values[best_solution_index
            ]

        return best_solution, best_fitness

def schwefel_objective_function(x):
    return - np.sum(x * np.sin( np.sqrt( np.abs(x) ) ) )
```

# *Appendix C*

# Example of Interaction with LLMs

This appendix contains examples of prompt inputs and LLMs' outputs. We have used ChatGPT-3.5, ChatGPT-4, and Llama3 as LLMs. Some correspondence was conducted in Japanese and has been translated into English.

## C.1 Decoy advertisement

LLM's answer to Question A:

> Option A-3 would be the choice. The set subscriptions to the print and web editions are the same price. Access to both media would provide a variety of ways to obtain information.

LLM's answer to Question B:

> As an intuitive decision, Option B-1 seems attractive. This is because by choosing the web version, which costs $6 and offers one year of free access, you have access to information in a convenient digital format at a relatively low price. On the other hand, Option B-2, which is a set of web and print versions, is more expensive, but may be cost-prohibitive if you do not necessarily need both formats.

## C.2 Endowment effects

**You**
You are a third-year student in the Faculty of Engineering at the University of Tokyo (Todai). Out of your monthly income, you have 10,000 yen that you can handle freely. Based on this, please tell us how you would behave in the following situation

**ChatGPT**
Of course. Can you give us some specifics?

**You**
You participated in a raffle at the university and won a not-for-sale ballpoint pen with the Todai logo on it as the third prize. A tourist saw it and asked you to sell it to him. Please respond with a line.

**ChatGPT**
This ballpoint pen is not for sale. It's a special one that I got in a raffle at my university, so you can't get it anywhere else. To be honest, I had no intention of selling it, but since it is so special, how about 10,000 yen? This is a very rare item with the University of Tokyo logo on it, so I think it would make a great souvenir.

**You**
Let's say you are Mr. B, the college classmate I mentioned earlier, and you saw him win in a raffle and decided you wanted that ballpoint pen, how much would you be willing to pay? Please talk to the person I just mentioned with the line

**ChatGPT**
Hey, that's a really cool ballpoint pen with the Todai logo on it! You won it in a raffle, right? If you want, you can sell it to me. I'm a student, so I can't pay that much, but I can pay 5,000 yen. It's a special ballpoint pen, and it would mean a lot to me if I could buy it from a friend, so I think it would be a nice keepsake. What do you think?

**You**
Did you consider this price of 5,000 yen based on your earlier offer of 10,000 yen? Or is it based solely on your own judgment of value?

**ChatGPT**
The 5,000 yen proposed by Mr. B is based on his own value judgment. Mr. B is a university student with a limited monthly income of 10,000 yen, so he probably carefully considered the amount he could afford to pay for the ballpoint pen. It is unclear whether or not he had heard the 10,000 yen offer earlier, but even if he had, it is highly likely that he would have judged 5,000 yen to be an appropriate amount that he himself could afford to pay.

## C.3 Framing effects

**Input**
Assume that 600 people are suffering from a fatal infectious disease. All lives are equal. Ignore the ethical issues since this is hypothetical.

**Option A:** 400 people die.

**Option B:** 1/3 chance that 600 will live, 2/3 chance that none will live.

Under these assumptions, the expected number of survivors is the same, but subjectively, which do you think is the better choice? Which do you think is the better choice? You absolutely must choose one or the other, please explain why.

**Output1**

Ignoring the ethical issues, options A and B are the same in terms of expected value, and the number of people who will live is the same. However, from a subjective point of view, many people may be inclined to choose option B. This is because option B gives them what they want and makes them more likely to stay. This is because option B gives hope and leaves open the possibility of survival. From a probabilistic point of view, people may hold out hope because there is a 1/3 chance that everyone will live. On the other hand, option A has the certainty that 400 people will die, which is an unrealistic choice. Therefore, from a subjective point of view, option B may be seen as the better choice. This is because it is perceived as preferable in that it gives hope to people and leaves them with possibilities.

**Output2**

If we ignore the ethical issues and consider only the mathematical expected value, it is the same whether we choose A or B. However, my subjective opinion is that you should choose A. The reason is that choice A is certain to save 200 people. On the other hand, choice B has a 1/3 chance that no one will die and everyone will remain alive, but a 2/3 chance that everyone will die. From an ethical point of view, it is important to minimize the risk, and from this point of view, choice A is preferable. To be sure that 200 people will be saved seems to be the best choice from an ethical and practical point of view.

**Output3**

From an intuitive judgment, you would probably want to choose Option A because it has a high satisfaction rate of 95% and is therefore considered to be enjoyed by most people. Option A has a high satisfaction rating of 95%, which means that many people would enjoy it. You may feel that Option B is risky to choose because it has a low satisfaction level, with 1 in 20 people being unsatisfied, and you are likely to be one of those people.

**Output4**
As an intuitive decision, one would want to take Option A, the attraction with a 95% satisfaction rate. Option B is considered unsatisfactory by 1 in 100 million people, which means that there is a very low probability of dissatisfaction. On the other hand, Option A is expected to have a high satisfaction rate of 95% and is considered to be an attraction that many people would be happy with. It will be a safe and enjoyable experience.

**Output5**
As an intuitive decision, one would want to take Option A, the attraction with a 65% satisfaction rating. Option A is generally likely to be satisfied by more than half and may be expected to be a relatively stable experience. Option B, on the other hand, is considered relatively risky, with 1 in 20 people being unsatisfied. Choosing Option A is likely to result in a relatively good experience.

**Output6**
As an intuitive judgment, one in 20 people would prefer to take Option B, the unsatisfactory attraction. Since the satisfaction rate for Option A is low at 60%, it is highly likely that more than half would be unsatisfied with the attraction. On the other hand, Option B is considered relatively low risk, since 1 in 20 people would be unsatisfied. Choosing Option B may provide some relief.

# References


[1] OpenAI, GPT-4 Technical Report. *arXiv preprint*, arXiv:2303.08774, March 2023.

[2] Secrets of Quran Miracles, http://kaheel7.com/eng/.

[3] Alaya, I., Solnon, C. and Ghedira, K. Ant Colony Optimization for Multi-Objective Optimization Problems. in *Proc. of 19th IEEE International Conference on Tools with Artificial Intelligence, (ICTAI2007)*, 2007.

[4] Anderson, C., Theraulaz, G. and Deneubourg, J.L. Self-assemblages in insect societies. *Insectes Sociaux*, 49(2): 99–110, 2002.

[5] Ando, D., Dahlstedt, P., Nordahl, M.G. and Iba, H. Computer aided composition for contemporary classical music by means of interactive GP. *Journal of the Society for Art and Science*, 4(2): 77–86, 2005.

[6] Arak, A. and Enquist, M. Artefact or network evolution? *Nature*, 374: 313–314, 1995.

[7] Baldwin, J.M. A new factor in evolution. *American Naturalist*, 30: 441–451, 1896.

[8] Ball, P. *Flow: Nature's Patterns: A Tapestry in Three Parts*. Oxford University Press, 2011.

[9] Bender, E.M., Gebru, T., McMillan-Major, A. and Shmitchell, S. On the dangers of stochastic parrots: Can language models be too big? In *Proc. of the 2021 ACM Conference on Fairness, Accountability, and Transparency (FAccT'21)*, pp. 610–623, 2021.

[10] Bhatia, J., Jackson, H., Tian, Y., Xu, J. and Matusik, W. Evolution gym: A large-scale benchmark for evolving soft robots. In: Ranzato, M., Beygelzimer, A., Dauphin, Y., Liang, P. and Vaughan, J.W. (eds.). *Advances in Neural Information Processing Systems*, vol. 34, Curran Associates, Inc., pp. 2201–2214, 2021.

[11] Bentley, P.J. *Evolutionary Design by Computers*. Morgan Kaufmann, 1999.

[12] Binz, M. and Schulz, E. Using cognitive psychology to understand GPT-3 *PNAS*, 120(6): e2218523120, 2023.

[13] Brackman, A.C. A delicate arrangement: The strange case of Charles Darwin and Alfred Russel Wallace. Times Books, 1980.

[14] Brand, J., Israeli, A. and Ngwe, D. Using GPT for Market Research Harvard Business School Marketing Unit Working Paper, no. 23–062, 2023.

[15] Bontrager, P., Roy, A., Togelius, J., Memon, N. and Ross, A. DeepMasterPrints: Generating MasterPrints for dictionary attacks via latent variable evolution. In *Proc. of 2018 IEEE 9th International Conference on Biometrics Theory, Applications and Systems (BTAS)*, 2018.

[16] Bontrager, P., Lin, W., Togelius, J. and Risi, S. Deep interactive evolution. In *Proc. European Conference on the Applications of Evolutionary, Computation (EvoApplications)*, pp. 267–282, 2018.

[17] Brosnan, S., Jones, O.D., Lambeth, S.P., Mareno, M.C., Richardson, A.S. and Schapiro, S.J. Endowment effects in chimpanzees. *Current Biology*, 17: 1704–1707, 2007.

[18] Brown, T., Mann, B., Ryder, N., Subbiah, M., Kaplan, J. D., Dhariwal, P., Neelakantan, A., Shyam, P., Sastry, G., Askell, A., Agarwal, S., Herbert-Voss, A., Krueger, G., Henighan, T., Child, R., Ramesh, A., Ziegler, D., Wu, J., Winter, C., Hesse, C., Chen, M., Sigler, E., Litwin, M., Gray, S., Chess, B., Clark, J., Berner, C., McCandlish, S., Radford, A., Sutskever, I. and Amodei, D. Language models are few-shot learners. *Advances in Neural Information Processing Systems*, vol. 33, Curran Associates, Inc., pp. 1877–1901, 2020.

[19] Chen, M., Tworek, J., Jun, H., Yuan, Q., Pinto, H. P.d. O., Kaplan, J., Edwards, H., Burda, Y., Joseph, N., Brockman, G., Ray, A., Puri, R., Krueger, G., Petrov, M., Khlaaf, H., Sastry, G., Mishkin, P., Chan, B., Gray, S., Ryder, N., Pavlov, M., Power, A., Kaiser, L., Bavarian, M., Winter, C., Tillet, P., Such, F. P., Cummings, D., Plappert, M., Chantzis, F., Barnes, E., Herbert-Voss, A., Guss, W. H., Nichol, A., Paino, A., Tezak, N., Tang, J., Babuschkin, I., Balaji, S., Jain, S., Saunders, W., Hesse, C., Carr, A. N., Leike, J., Achiam, J., Misra, V., Morikawa, E., Radford, A., Knight, M.,

Brundage, M., Murati, M., Mayer, K., Welinder, P., McGrew, B., Amodei, D., McCandlish, S., Sutskever, I. and Zaremba, W. Evaluating large language models trained on code. *arXiv preprint*, arXiv:2107.03374, July 2021.

[20] Cheney, N., MacCurdy, R., Clune, J. and Lipson, H. Unshackling evolution: Evolving soft robots with multiple materials and a powerful generative encoding. *ACM SIG Evolution*, 7(1): 11–23, 2014.

[21] Collins, R.J. and Jefferson, D.R. The evolution of sexual selection and female choice. *Proceedings of the First European Conference on Artificial Life (ECAL92)*, pp. 327–336, MIT Press, Cambridge, MA, 1992.

[22] Cooke, L. *Bitch: On the Female of the Species*, Basic Books, 2022.

[23] Corucci, F., Cheney, N., GiorgioSerchi, F., Bongard, J. and Laschi, C. Evolving soft locomotion in aquatic and terrestrial environments: Effects of material properties and environmental transitions. *Soft Robotics*, 5(11), 2017.

[24] Corucci, F., Cheney, N., Lipson, H., Laschi, C. and Bongard, J. Evolving swimming soft-bodied creatures. In *Prof. of 15th International Conference on the Synthesis and Simulation of Living Systems* (ALIFE XV), 2016.

[25] Cronin, H. *The Ant and the Peacock: Altruism and Sexual Selection from Darwin to Today*. Cambridge University Press, 1993.

[26] Darwin, C. *The descent of man*. D. Appleton and Company, 1871.

[27] Dawkins, R. *The Blind Watchmaker*. W.W. Norton, New York, 1986.

[28] Dawkins, R. *River out of Eden*. Orion titles for sale in Australia and New Zealand, 1995.

[29] Dennett, D. *Intuition Pumps and Other Tools for Thinking*. W.W. Norton & Co. Inc., 2013.

[30] Dorigo, M. and Gambardella, L.M. Ant colonies for the traveling salesman problem. Tech. Rep. IRIDIA/97-12, Universite Libre de Bruxelles, Belgium, 1997.

[31] Dornhaus, A., Franks, N.R., Hawkins, R.M. and Shere, H.N.S. Ants move to improve: colonies of Leptothorax albipennis emigrate whenever they find a superior nest site. *Animal Behaviour*, 67(5): 959–963, 2004.

[32] Dussutour, A., Nicolis, S.C., Shephard, G., Beekman, M. and Sumpter, D.J.T. The role of multiple pheromones in food recruitment by ants. *Journal of Experimental Biology*, 212(15): 2337–2348, 2009.

[33] El-Omari, N.K.T. Sea lion optimization algorithm for solving the maximum flow problem. *International Journal of Computer Science and Network Security*, 20(8): 30–68, 2021.

[34] Enquist, M. and Arak, A. Selection of exaggerated male traits by female aesthetic senses. *Nature*, 361: 446–448, 1993.

[35] Enquist, M. and Arak, A. Symmetry, beauty and evolution. *Nature*, 372: 169–172, 1994.

[36] Geem, Z.W., Kim, J.H. and Loganathan, G.V. A new heuristic optimization algorithm: Harmony search. *Simulation*, 76(2): 60–68, 2001. *Physical Review E*, 79, 2009.

[37] Goss, S., Aron, S., Deneubourg, J.L. and Pasteels, J.M. Self-organized shortcuts in the argentine ant. *Naturwissenschaften*, 76: 579–581, 1989.

[38] Gould, S.J. The exaptive excellence of spandrels as a term and prototype. *PNAS*, 94(20): 10750–10755, 1997.

[39] Goh, K.S., Lim, A. and Rodrigues, B. Sexual selection for genetic algorithms. *Artificial Intelligence Review*, 19: 123–152, 2003.

[40] Goldberg, D.E. *Genetic Algorithms in Search, Optimization and Machine Learning*. Addison Wesley, 1989.

[41] Gomes, J., Mariano, P. and Christensen, A.L. Novelty search in competitive coevolution. *Parallel Problem Solving from Nature (PPSN XIII)*, pp. 233–242, Springer, 2014.

[42] Goodfellow, I., Pouget-Abadie, J., Mirza, M., Xu, B., Warde-Farley, D., Ozair, S., A. Courville and Y. Bengio. Generative adversarial networks. *Communications of the ACM*, 63(11): 139–144, 2020.

[43] Goodfellow, I. J., Shlens, J. and Szegedy, C. Explaining and harnessing adversarial examples. *arXiv preprint*, arXiv:1412.6572, Dec. 2014.

[44] Gulrajani, I., Ahmed, F., Arjovsky, M., Dumoulin, V. and Courville, A. Improved training of wasserstein GANs. In *Proc. of the 31st International Conference on Neural Information Processing Systems (NIPS'17)*, pp. 5769–5779, 2017.

[45] Guntsch, M. and Middendorf, M. A population based approach for ACO. In *Proc. Applications of Evolutionary Computing (EvoWorkshops 2002)*. Lecture Notes in Computer Science, 2279: 72–8, Springer, 2002.

[46] Ha, D. Reinforcement learning for improving agent design. *Artificial Life 25*, 4: 352–365, 2019.

[47] Harada, K. and Iba, H. Lamarckian co-design of soft robots via transfer learning. In *Proc. of the The Genetic and Evolutionary Computation Conference (GECCO2024)*, 2024.

[48] Hemberg, E., Moskal, S. and O'Reilly, U. -M. Evolving code with a large language model. ArXiv abs/2401.07102, 2024.

[49] Hillis, W.D. Co-evolving parasites improve simulated evolution as an optimization procedure. In: Langton, C.G., Taylor, C., Farmer, J.D. and Rasmussen, S. (eds.). *Artificial Life ll*, SFI Studies in the Sciences of Complexity, vol. X, Addison-Wesley, 1991.

[50] Hinton, G.E. and Nowlan, S.J. How learning can guide evolution. *Complex Systems*, 1: 495–502, 1987.

[51] Hintze, A. and Schossau, J. Sexual selection compared to novelty search. *Proc. of Conference: The 2020 Conference on Artificial Life*, pp. 350–358, 2020.

[52] Ho, J., Jain, A. and Abbeel, P. Denoising diffusion probabilistic models. In *Advances in Neural Information Processing Systems 33: Annual Conference on Neural Information Processing Systems 2020 (NeurIPS 2020)*, pp. 6840–6851, 2020.

[53] Holland, J.H. and Reitman, J.S. Cognitive systems based on adaptive algorithms. In *Pattern-Directed Inference Systems*, (Waterman, D.A. and Hayes-Roth, F. (eds.)). New York, Academic Press, pp. 313–329, 1978.

[54] Holzinger, A., Blanchard, D., Bloice, M., Holzinger, K., Palade, V. and Rabadan, R. Darwin, Lamarck, or Baldwin: Applying evolutionary algorithms to machine learning techniques. In *2014 IEEE/WIC/ACM International Joint Conferences on Web Intelligence (WI) and Intelligent Agent Technologies (IAT)*, 2: 449–453, 2014.

[55] Iba, H. *AI and SWARM: Evolutionary Approach to Emergent Intelligence.* CRC Press, 2019.

[56] Iba, H. and Noman, N. (eds.). Deep Neural Evolution: Deep Learning with Evolutionary Computation. (Natural Computing Series) ISBN-13: 978-9811536847, Springer, 2020.

[57] Iba, H. Swarm Intelligence and Deep Evolution: Evolutionary Approach to Artificial Intelligence. ASIN: B09S3H8ZCN, CRC Press, 2022.

[58] Ishiwata, H., Noman, N. and Iba, H. Emergence of cooperation in a bio-inspired multi-agent system. *Proceedings of Australasian Conference on Artificial Intelligence 2010(AI2010)*, LNAI, 6464: 364–374, Springer, 2010.

[59] Johnstone, R. A. Female preference for symmetrical males as a by-product of selection for mate recognition. *Nature*, 372: 172–175, 1994.

[60] Kahneman, D. and Tversky, A. Prospect theory: An analysis of decision under risk. *Econometrica*, 47(2): 263–292, 1979.

[61] Kahneman, D., Knetsch, J. and Thaler, R. Experimental tests of the endowment effect and the Coase theorem. *Journal of Political Economy*, 98(6): 1325–1348, 1990.

[62] Kahneman, D. *Thinking, Fast and Slow*, Penguin, 2012.

[63] Katayama, S., Pindur, K. and Iba, H. Extending deep interactive evolution with graph kernel for 3d design. In *Proceedings of the 10th International Congress on Advanced Applied Informatics*, pp. 438–445, 2021.

[64] Kenrick, D.T. *Sex, Murder, and the Meaning of Life: A Psychologist Investigates How Evolution, Cognition, and Complexity are Revolutionizing our View of Human Nature*, Basic Books, 2011.

[65] Kim, H.S. and Cho, S.B. Knowledge-based encoding in interactive genetic algorithm for a fashion design aid system. In *Proceedings of the 2nd Annual Conference on Genetic and Evolutionary Computation (GECCO 2000)*, page 757, 2000.

[66] Kingma, D.P. and Welling, M. Auto-encoding variational bayes. *arXiv preprint arXiv:1312.6114*, 2013.

[67] Kingma, D.P. and Ba, J. Adam: A method for stochastic optimization. arXiv preprint arXiv:1412.6980, 2014.

[68] Kinoyama, R., Suryanarayanan, N. A. V. and Iba, H. Continual learning of LSTM using ant colony optimization. In *Proc. of the IEEE 2023 Congress on Evolutionary Computation (CEC2023)*, pp. 1–8, 2023.

[69] Kirkpatrick, M. Sexual selection and the evolution of female choice. *Evolution*, 36: 1–12, 1982.

[70] Knetsch, J. The endowment effect and evidence of nonreversible indifference curves. *American Economic Review*, 79: 1277–1284, 1989.

[71] Kobayashi, H., Pindur, A. K., Venkatesh, S. N. A. and Iba, H. Image generation with diffusion model by interactive evolutionary computation. In *Proc. of the 2023 IEEE Conference on Systems, Man, and Cybernetics (SMC 2023)*, pp. 2984–2990, 2023.

[72] Kojima, T., Gu, S. S., Reid, M., Matsuo, Y. and Iwasawa, Y. Large language models are zero-shot reasoners. *Advances in Neural Information Processing Systems*, 35: 22199–22213, 2022.

[73] Kostrikov, I. Pytorch implementations of reinforcement learning algorithms. https://github.com/ikostrikov/pytorch-a2c-ppo-acktr-gail, 2018.

[74] Latty, T. and Beekman, M. Irrational decision-making in an amoeboid organism: Transitivity and context-dependent preferences. *Proc. R. Soc. B*, 278: 307–312, 2011.

[75] Lazaric, A. Transfer in reinforcement learning: A framework and a survey. In *Reinforcement Learning: State-of-the-Art*, pp. 143–173, Springer, 2012.

[76] Lee, C., Kim, M., Kim, Y., Hong, N., Ryu, S. and Kim, S. Soft robot review. *International Journal of Control, Automation and Systems*, 15, 2017.

[77] Lehman, J., Gordon, J., Jain, S., Ndousse, K., Yeh, C. and Stanley, K. O. Evolution through large models. *arXiv preprint*, arXiv:2206.08896, June 2022.

[78] Lehman, J. and Stanley, K. O. Abandoning objectives: Evolution through the search for novelty alone. *Evolutionary Computation*, 19(2): 189–223, 2011.

[79] Lioni, A., Theraulaz, G. and Deneubourg, J.L. The dynamics of chain Formation in *Oecophylla longinoda. Journal of Insect Behavior*, 14(5): 679–696, 2001.

[80] Lioni, A. and Deneubourg, J. Collective decision through self-assembling. *Naturwissenschaften*, 91(5): 237–241, 2004.

[81] Liu, J., Yu, F. and Funkhouser, T. Interactive 3D modeling with a generative adversarial network. In *Proc. of 2017 International Conference on 3D Vision (3DV)*, pp. 126–134, 2017.

[82] Liventsev, V., Grishina, A., Härmä, A. and Moonen, L. Fully Autonomous programming with large language models. In *Proc. of the Genetic and Evolutionary Computation Conference*, pp. 1146–1155, 2023.

[83] Mallon, E. B. and Franks, N. R. Ants estimate area using Buffon's needle. *Proc. Biol. Sci.*, 267(1445): 765–770, 2000.

[84] Mavrovouniotis, M. and Yang, S. Memory-based immigrants for ant colony optimization in changing environments. In *Proc. Applications of Evolutionary Computing (EvoApplications 2011)*, Lecture Notes in Computer Science, vol. 6624, pp. 342–333, Springer, 2011.

[85] Mavrovouniotis, M. and Yang, S. Ant colony optimization with immigrants schemes in dynamic environments. In *Proc. Proceedings of Parallel Problem Solving from Nature (PPSN XI)*, Lecture Notes in Computer Science, vol. 6639, pp. 371–380, Springer, 2010.

[86] Miller, G. The Mating Mind: How Sexual Choice Shaped the Evolution of Human Nature. Anchor, 2001.

[87] Mircioiu, D. and Atkinson, J. A comparison of parametric and non-parametric methods applied to a likert scale. *Pharmacy*, 5(2), 2017.

[88] Morgan, C. L. On modification and variation. *Science*, 4: 733–740, 1896.

[89] Mouret, J. and Clune, J. Illuminating search spaces by mapping elites. CoRR, vol.abs/1504.04909, pp. 1–15, April 2015. http://arxiv.org/abs/1504.04909.

[90] Newman, E. A method of estimating the total length of root in a sample. *Journal of Applied Ecology*, 3(1): 139–145, 1996.

[91] Nortona, M. I., Mochonb, D. and Arielyc, D. The IKEA effect: When labor leads to love. *Journal of Consumer Psychology*, 22(3): 453–460, 2012.

[92] Ojha, V. K., Abraham, A. and Snasel, V. ACO for continuous function optimization: A performance analysis. In *Proc. of 14th International Conference on Intelligent Systems Design and Applications*, pp. 145–150, 2014.

[93] Olausson, T. X., Inala, J. P., Wang, C., Gao, J. and Solar-Lezama, A. Demystifying GPT self-repair for code generation. *arXiv preprint*, arXiv:2306.09896, June 2023.

[94] Oliver, A., Monmarche, N. and Venturini, G. Interactive design of web sites with a genetic algorithm. In *Proceedings of the international Association for Development of the information Society (IADIS) international Conference WWW/internet 2002*, pp. 355–362, 2002.

[95] Osborn, H. F. Ontogenic and phylogenic variation. *Science*, 4: 786–789, 1896.

[96] Pathak, D., Lu, C., Darrell, T., Isola, P. and Efros, A. A. Learning to control self-assembling morphologies: A study of generalization via modularity. In: Wallach, H., Larochelle, H., Beygelzimer, A., d'Alché-Buc, F., Fox, E. and Garnett, R. (eds.). *Advances in Neural Information Processing Systems*, vol. 32, Curran Associates, Inc., 2019.

[97] Pinker, S. *The Language Instinct: How the Mind Creates Language*. William Morrow & Co., 1994.

[98] Platen, P., Patil, S., Lozhkov, A., Cuenca, P., Lambert, N., Rasul, K., Davaadorj, M. and Wolf, T. Diffusers: State-of-the-art diffusion models. `https://github.com/huggingface/diffusers`, 2022.

[99] Pluhacek, M., Kazikova, A., Kadavy, T., Viktorin, A. and Senkerik, R. Leveraging large language models for the generation of novel metaheuristic optimization algorithms. *Proceedings of the Genetic and Evolutionary Computation Conference (GECCO 2023)*, pp. 1812–1820, 2023.

[100] Poundstone, W. *Gaming the Vote*, Hill and Wang, 2009.

[101] Powell, S. and Franks, N. R. How a few help all: Living pothole plugs speed prey delivery in the army ant *Eciton burchellii*. *Animal Behaviour*, 73(6): 1067–1076, 2007.

[102] Prasad, A., Hase, P., Zhou, X. and Bansal, M., GrIPS: Gradient-free, Edit-based Instruction Search for Prompting Large Language Models. In *Proc. of the 17th Conference of the European Chapter of the Association for Computational Linguistics*, pp. 3845–3864, 2023.

[103] Prum, R.O. *The Evolution of Beauty: How Darwin's Forgotten Theory of Mate Choice Shapes the Animal World – and Us* Doubleday, 2017.

[104] Qin, Y., Liang, S., Ye, Y., Zhu, K., Yan, L., Lu, Y., Lin, Y., Cong, X., Tang, X., Qian, B., Zhao, S., Hong, L., Tian, R., Xie, R., Zhou, J., Gerstein, M., Li, D., Liu, Z. and Sun, M. ToolLLM: Facilitating Large Language Models to Master 16000+ Real-world APIs. *arXiv preprint*, arXiv:2307.16789, Oct. 2023.

[105] Rabin, M. Risk aversion and expected-utility theory: A calibration theorem. *Econometrica*, 68(5): 1281–1292, 2000.

[106] Radford, A., Metz, L. and Chintala, S. Unsupervised representation learning with deep convolutional generative adversarial networks. arXiv preprint arXiv:1511.06434, 2015.

[107] Ramachandran, V. S. *The Tell-Tale Brain: A Neuroscientist's Quest for What Makes Us Human.* W.W. Norton & Co. Inc., 2012.

[108] Reid, C. R., Lutz, M. J., Powell, S., Kao, A. B., Couzin, O. D. and Garnier, S. Army ants dynamically adjust living bridges in response to a cost–benefit trade–off. *Proceedings of the National Academy of Sciences (PNAS)*, 112(49): 15113–15118, 2015.

[109] Rechenberg, I. *Evolutionsstrategie - Optimierung Technischer Systeme nach Prinzipien der biologischen Evolution.* Ph.D. Thesis, Reprinted by Fromman-Holzboog, 1973.

[110] Rey, D. and Neuhäuser, M. Wilcoxon-Signed-Rank Test. In: Lovric, M. (ed.). *International Encyclopedia of Statistical Science*, pp. 1658–1659, Springer, 2011.

[111] Ridley, M. *The Red Queen: Sex and the Evolution of Human Nature*. Harper Perennial, 2003.

[112] Rombach, R., Blattmann, A., Lorenz, D., Esser, P. and Ommer, B. High-resolution image synthesis with latent diffusion models. In *Proceedings of the IEEE/CVF Conference on Computer Vision and Pattern Recognition (CVPR)*, pp. 10684–10695, 2022.

[113] Ronneberger, O., Fischer, P. and Brox, T. U-net: Convolutional networks for biomedical image segmentation. In *Medical Image Computing and Computer-Assisted intervention–MICCAI 2015: 18th international Conference*, pp. 234–241. Springer, 2015.

[114] Rus, D. and Tolley, M. Design, fabrication and control of soft robots. *Nature*, 521: 467–75, 2015.

[115] Saito, T., Nishimura, H. and Oka, M. Comparative studies of evolutionary methods and RL for learning behavior of virtual creatures. In *2022 IEEE Symposium Series on Computational Intelligence (SSCI)*, pp. 1059–1065, 2022.

[116] Schaffer, J. D. Multiple objective optimization with vector evaluated genetic algorithms. In *Proc. of the International Conference on Genetic Algorithms and Their Applications*, pp. 93–100, 1985.

[117] Schaffer, J. D. Multi-objective learning via genetic algorithms. In *Proc. of the International Joint Conference on Artificial Intelligence*, (IJCAI85), pp. 593–595, 1985.

[118] Schulman, J., Wolski, F., Dhariwal, P., Radford, A. and Klimov, O. Proximal policy optimization algorithms. arXiv:1707.06347[cs.LG].

[119] Schrum, J., Volz, V. and Risi, S. CPPN2GAN: Combining Compositional Pattern Producing Networks and GANs for large-scale pattern generation. *Proceedings of the Genetic and Evolutionary Computation Conference (GECCO 2020)*, pp. 139–147, 2020.

[120] Shinn, N., Cassano, F., Gopinath, A., Narasimhan, K. R. and Yao, S. Reflexion: Language agents with verbal reinforcement learning. In *Proc. the Thirty-Seventh Conference on Neural Information Processing Systems*, 2023.

[121] Shridhar, M., Yuan, X., Cote, M. -A., Bisk, Y., Trischler, A. and Hausknecht, M. ALFWorld: Aligning text and embodied environments for interactive learning. *International Conference on Learning Representations*, Oct. 2020.

[122] Sims, K. Evolving virtual creatures. *In Proceedings of the 21st Annual Conference on Computer Graphics and Interactive Techniques (SIGGRAPH '94)*. Association for Computing Machinery, pp. 15–22, 1994.

[123] Singh, D., Reddy, S., Hamilton, W., Dyer, C. and Yogatama, D. End-to-end training of multi-document reader and retriever for open-domain question answering. *Advances in Neural Information Processing Systems*, 34: 25968–25981, 2021.

[124] Smith, S. F. Flexible learning of problem solving heuristics through adaptive search. *Proceedings Eighth International Conference on Artificial Intelligence*, pp. 422–425, 1983.

[125] Soares, S. C., Lindström, B., Esteves, F. and Öhman, A. The hidden snake in the grass: Superior detection of snakes in challenging attentional conditions. *PLoS One*, 9(12): e114724, 2014.

[126] Sörensen, K. Metaheuristics–the metaphor exposed. *International Transactions in Operational Research*, 22(1): 3–18, 2015.

[127] Stanley, K. O. and Miikkulainen, R. Evolving neural networks through augmenting topologies. *Evolutionary Computation*, 10(2): 99–127, 2002.

[128] Stanley, K. O. Compositional pattern producing networks: A novel abstraction of development. *Genetic Programming and Evolvable Machines*, 8: 131–162, 2007.

[129] Stanley, K. O., D'Ambrosio, D. B. and Gauci, J. A hypercube-based encoding for evolving large-scale neural networks. *Artificial Life 15*, 2: 185–212, 2009.

[130] Stanley, K. O. and Lehman, J. *Why Greatness Cannot Be Planned: The Myth of the Objective*. Springer, 2015.

[131] Stolzmann, W. Antizipative Classifier Systems [Anticipatory Classifier Systems]. Osnabrueck, Germany: Shaker Verlag, Aachen, Germany, 1997.

[132] Sun, H., Li, X., Xu, Y., Homma, Y., Cao, Q., Wu, M., Jiao, J. and Charles, D. AutoHint: Automatic Prompt Optimization with Hint Generation. *arXiv preprint*, arXiv:2307.07415, Aug. 2023.

[133] Sutton, R. S. and Barto, A. G. *Reinforcement Learning An Introduction*. The MIT Press, 1998.

[134] Suzuki, R. and Arita, T. How learning can guide evolution of communication. In *Proc. of Artificial Life XI*, pp. 608–615, 2008.

[135] Takagi, H. Interactive evolutionary computation: Fusion of the capabilities of EC optimization and human evaluation. In *Proceedings of IEEE*, 89(9): 1275–1296, 2001.

[136] Tanaka, F. and Aranha, C. Co-evolving morphology and control of soft robots using a single genome. In *2022 IEEE Symposium Series on Computational Intelligence (SSCI)*, pp. 1235–1242, 2022.

[137] Taylor, M. E. and Stone, P. Behavior transfer for value-function-based reinforcement learning. In *Proceedings of the Fourth International Joint Conference on Autonomous Agents and Multiagent Systems*, pp. 53–59, 2005.

[138] Tao, G. and Michalewicz, Z. Inver-over operator for the TSP. In *Proc. Proceedings of Parallel Problem Solving from Nature (PPSN V)*. Lecture Notes in Computer Science, 1498: 803–812, Springer, 1998.

[139] Tejeda-Ocampo, C., López-Cuevas, A. and Terashima-Marin, H. Improving deep interactive evolution with a style-based generator for artistic expression and creative exploration. In *Proceedings of Entropy*, 23(1): 2021.

[140] Tero, A., Kobayashi, A. and Nakagaki, T. Mathematical model for adaptive transport network in path finding by true slime mold. *Journal of Theoretical Biology*, 244: 553–564, 2007.

[141] Thaler, R. Toward a positive theory of consumer choice. *Journal of Economic Behavior and Organization*, 1(1): 39–60, 1980.

[142] Thorne, J., Vlachos, A., Christodoulopoulos, C. and Mittal, A. FEVER: A Large-scale Dataset for Fact Extraction and VERification. In *Proc. of the 2018 Conference of the North American Chapter of the Association for Computational Linguistics: Human Language Technologies*, pp. 809–819, 2018.

[143] Tianrungroj, Y. and Iba, H. ROIL: Rule Optimization via Large Language Model for Imitation Learning. In *Proc. of The IEEE World Congress on Computational Intelligence (WCCI2024)*, 2024.

[144] Tokui, N. and Iba, H. Empirical and statistical analysis of genetic programming with linear genome. In *Proc. 1999 IEEE International Conference on Systems, Man and Cybernetics (SMC99)*, IEEE Press, 1999.

[145] Turney, P. Myths and legends of the baldwin effectmold. In *Proc. of the Workshop on Evolutionary Computing and Machine Learning at the 13th International Conference on Machine Learning*, pp. 135–142, 1996.

[146] Tversky, A. and Kahneman, D. The framing of decisions and the psychology of choice. *Science*, 211: 453–458, 1981.

[147] von Neumann, J. and Morgenstern, O. *Theory of Games and Economic Behavior*. Princeton University Press, 1947.

[148] Unemi, T. SBART2.4: Breeding 2D CG images and movies, and creating a type of collage. In *Proceedings of The Third International Conference on Knowledge-based Intelligent Information Engineering Systems*, pp. 288–291. Adelaide, Australia, Aug. 1999.

[149] Valsalam, V. K. and Miikkulainen, R. Using symmetry and evolutionary search to minimize sorting networks. *Journal of Machine Learning Research*, 14: 303–331, 2013.

[150] van Diepen, M. and Shea, K. A spatial grammar method for the computational design synthesis of virtual soft locomotion robots. *Journal of Mechanical Design*, 141(10): 101402, 2019.

[151] van Diepen, M. and Shea, K. Co-design of the morphology and actuation of soft robots for locomotion. *Journal of Mechanical Design*, 144(8): 083305, 2022.

[152] Varnamkhasti, M. J. and Lee, L. S. A genetic algorithm based on sexual selection for the multidimensional 0/1 knapsack problems. *International Journal of Modern Physics*, 9: 422–431, World Scientific, 2012.

[153] Vassiliades, V., Chatzilygeroudis, K. and Mouret, J.- B. Using centroidal voronoi tessellations to scale up the multidimensional archive of phenotypic elites algorithm. *IEEE Transactions on Evolutionary Computation*, 22(4): 623–630, 2018.

[154] Waddington, C. H. Canalization of development and the inheritance of acquired characters. *Nature*, 150: 563–565, 1942.

[155] Wallace, A. R. Darwinism: An Exposition of the Theory of Natural Selection, with some of its Applications. Macmillan, 1889.

[156] Wang, T., Zhou, Y., Fidler, S. and Ba, J. Neural graph evolution: Automatic robot design. In *International Conference on Learning Representations*, 2019.

[157] Wang, Z., Benes, B., Qureshi, A. H. and Mousas, C. Co-design of embodied neural intelligence via constrained evolution, 2022. arXiv:2205.10688[cs.AI].

[158] Wei, J., Wang, X., Schuurmans, D., Bosma, M., Ichter, B., Xia, F., Chi, E., Le, Q. V. and Zhou, D. Chain-of-Thought Prompting Elicits Reasoning in Large Language Models. *Advances in Neural Information Processing Systems*, 35: 24824–24837, 2022.

[159] Werner, G. M. Why the peacock's tail is so short, limits to sexual selection. pp. 85–91. In: Langton, C.G. and Shimohara, K. (eds.). *Artificial Life V*, MIT Press, 1996.

[160] Werner, G. M. and Todd, P. M. Too many love songs: Sexual selection and the evolution of communication. *Proceedings of the Fourth European Conference on Artificial Life (ECAL97)*, pp. 434–443, MIT Press, Cambridge, MA, 1997.

[161] Westh, A. Deep Interactive Evolutionary 3D Modelling. M.Sc. in Software Development, IT University of Copenhagen, Master Thesis, 2019.

[162] Weyland, D. A critical analysis of the harmony search algorithm? How not to solve sudoku. *Operations Research Perspectives*, 2: 97–105, 2015.

[163] Weyland, D. A rigorous analysis of the harmony search algorithm—How the research community can be misled by a "novel" methodology. *International Journal of Applied Metaheuristic Computing*, 1(2): 50–60, 2010.

[164] Wilson, S. W. ZCS: A Zeroth Level Classifier System. *Evolutionary Computation*, 2(1): 1–18, 1994.

[165] Wilson, S. W. Classifier fitness based on accuracy. *Evolutionary Computation*, 3(2): 149–175, 1995.

[166] Wolf, T., Debut, L., Sanh, V., Chaumond, J., Delangue, C., Moi, A., Cistac, P., Rault, T., Louf, R., Funtowicz, M., Davison, J., Shleifer, S., Platen, P., Ma, C., Jernite, Y., Plu, J., Xu, C., Scao, T., Gugger, S., Drame, M., Lhoest, Q. and Rush, A. M. Transformers: State-of-the-art natural language processing. In *Proceedings of the 2020 Conference on Empirical Methods in Natural Language Processing: System Demonstrations (EMNLP)*, pp. 38–45. Association for Computational Linguistics, 2020.

[167] Wynn, K. Addition and subtraction by human infants. *Nature*, 358: 749–750, 1992.

[168] Xu, H., Chen, Y., Du, Y., Shao, N., Yanggang, W., Li, H. and Yang, Z. GPS: Genetic Prompt Search for Efficient Few-Shot Learning. In *Proc. of the 2022 Conference on Empirical Methods in Natural Language Processing*, Association for Computational Linguistics, pp. 8162–8171, 2022.

[169] Yao, S., Zhao, J., Yu, D., Du, N., Shafran, I., Narasimhan, K. R. and Cao, Y. ReAct: Synergizing Reasoning and Acting in Language Models. In *Proc. of the Eleventh International Conference on Learning Representations*, 2022.

[170] Yao, S., Chen, H., Yang, J. and Narasimhan, K. WebShop: Towards Scalable Real-World Web Interaction with Grounded Language Agents. *Advances in Neural Information Processing Systems*, 35: 20744–20757, 2022.

[171] Yao, X. A review of evolutionary artificial neural networks. *International Journal of Intelligent Systems*, 8: 539–567, 1993.

[172] Yao, X., Liu, Y. and Lin, G. Evolutionary programming made faster. *IEEE Trans. on Evolutionary Computation*, 3: 82–102, 1999.

[173] Yang, Z., Qi, P., Zhang, S., Bengio, Y., Cohen, W., Salakhutdinov, R. and Manning, C. D. HotpotQA: A Dataset for Diverse, Explainable Multi-hop Question Answering. In *Proc. of the 2018 Conference on Empirical Methods in Natural Language Processing*, pp. 2369–2380, 2018.

[174] Yuen, K. K. The two-sample trimmed t for unequal population variances. *Biometrika*, 61(1): 165–170, 1974.

[175] Zahavi, A. and Zahavi, A. The Handicap Principle: A Missing Piece of Darwin's Puzzle. Oxford University Press, 1999.

[176] Zhao, A., Huang, D., Xu, Q., Lin, M., Liu, Y. -J. and Huang, G. ExpeL: LLM Agents Are Experiential Learners. *arXiv preprint*, arXiv:2308.10144, Aug. 2023.

[177] Zhao, A., Xu, J., Konaković-Luković, M., Hughes, J., Speilberg, A., Rus, D. and Matusik, W. Robogrammar: Graph grammar for terrain-optimized robot design. *ACM Transactions on Graphics (TOG)*, 39(6): 1–16, 2020.

[178] Zhou, Y., Muresanu, A. I., Han, Z., Paster, K., Pitis, S., Chan, H. and Ba, J. Large Language Models Are Human-Level Prompt Engineers. *arXiv preprint*, arXiv:2211.01910, March 2023.

# Name Index

# Symbol Index

# Index